Quantum Chemistry and Computing for the Curious

Illustrated with Python and Qiskit® code

Keeper L. Sharkey

Alain Chancé

BIRMINGHAM—MUMBAI

Quantum Chemistry and Computing for the Curious

Copyright © 2022 Packt Publishing

Group Product Manager: Richa Tripathi
Publishing Product Manager: Richa Tripathi
Senior Editor: Matthew Moodie
Technical Editor: Pradeep Sahu
Copy Editor: Safis Editing
Project Coordinator: Manisha Singh
Proofreader: Safis Editing
Indexer: Tejal Daruwale Soni
Production Designer: Roshan Kawale
Marketing Coordinator: Pooja Yadav

First published: May 2022
Production reference: 1120522

Published by Packt Publishing Ltd.
Livery Place
35 Livery Street
Birmingham
B3 2PB, UK.

ISBN 978-1-80324-390-0

www.packt.com

To my mother, Karen, for always encouraging me to pursue my love of chemistry and who unexpectedly passed away during the writing process of this book.

- Keeper L. Sharkey

To Elaine, my wife, this book is dedicated with love.

- Alain Chancé

Foreword

I am honored to write this foreword to *Quantum Chemistry and Computing for the Curious.*

I met Keeper in 2019 during the Quantum.Tech Congress in Boston, where she was presenting her ideas on quantum chemistry. It was clear that she was the leading expert in this field and stayed in touch with her ever since. I have worked in various areas of quantum computing with mostly financial use cases; however, I have been exposed to some quantum machine learning and quantum chemistry use cases as well. My general sense has been that the literature and examples that lay out how quantum computing applies to chemistry use cases are sorely lacking.

In 2021, during a conversation, I encouraged Keeper to write this book for the benefit of the quantum computing ecosystem, and it turned out she had already been thinking of it and agreed. I was even more excited to hear that Alain, who is well known in the Qiskit community, had agreed to co-author the book with her, knowing that it would add to the useability and make the topic much more approachable to those with some Qiskit background.

If you are reading this, clearly you have an interest in quantum computing, but also in molecule simulations and quantum chemistry use cases. Richard Feynman wrote that *"Nature isn't classical, dammit, and if you want to make a simulation of nature, you'd better make it quantum mechanical, and by golly it's a wonderful problem, because it doesn't look so easy."* All the points of this statement are true and beautifully come out in this book.

Most people in quantum computing are aware that we use the Variational Quantum Eigensolver (VQE) to efficiently obtain the potential energy surfaces (PES) for small molecules. However, applying this approach requires understanding classical methods that may or may not be variational, quantum mechanics, developing the energy equation or Hamiltonian of the molecule, making very specific assumptions, such as the Born-Oppenheimer (BO) approximation, and then converting the Hamiltonian into a quantum circuit before it can be solved using VQE. The Hartree-Fock (HF) theory is used to describe the motion of each electron by a molecular orbital, which in turn is made up of a linear set of atom-centered basis functions. STO-3G is one example. Various techniques, including Bravyi-Kitaev (BK) and Jordan-Wigner (JW) are used to transform a chemical Hamiltonian into a qubit Hamiltonian. We need to ensure that these transformations use qubit operators that represent the fermionic operators. We also need to ensure certain symmetries related to the number of electrons, spin, and time-reversal are preserved during this transformation. This is the convergence of many disciplines and where things get complicated very quickly.

Keeper and Alain lead the reader step-by-step through the many topics, techniques, and fundamentals with detailed explanations and code samples. The book introduces quantum concepts such as the structure of light and the atom and then dives into the required areas of quantum mechanics, such as electron orbital structure, the Pauli exclusion principle (PEP), and Schrödinger's equations. Next, the book introduces key quantum computing concepts, including the Bloch sphere, quantum gates, Bell states, the density matrix, and symmetrized versus anti-symmetrized states. We then dive into the core of quantum simulation with the BO approximation, Fock space, creation and annihilation operators, basis sets, and fermionic to qubit mappings, before introducing the reader to Qiskit Nature as a tool to easily apply the above principles. There are detailed explanations, and the equations and concepts are carefully developed and expanded in far more detail than even I would have imagined. In the hero's journey, all our intellectual faculties are challenged as we finally reach our goal as we begin to put all the concepts that we have meticulously mastered through the VQE algorithm and find the ground state of three different molecules on a quantum computer.

In 2019, Keeper mentioned to me that one day, quantum computers could provide more accurate simulations of complex molecules, where approximations required for classical computations would not be needed. Is this the future for quantum simulations? You will find out more as Keeper takes us beyond Born-Oppenheimer.

This book is a goldmine for those wanting the breadth and depth of information required to understand quantum simulations in one place. Keeper and Alain have made a sizable and worthy contribution to the growing wealth of quantum computing literature. I think Richard Feynman would be proud.

Alex Khan

Entrepreneur, advisor, and educator in quantum computing

Baltimore, MD

March 2022

Contributors

About the authors

Keeper L. Sharkey, PhD is the founder and CEO of ODE, L3C, a social enterprise that serves through Quantum Science, Technology and Research, qSTAR. She is Chair of Quantum Applied Chemistry at the Quantum Security Alliance. She obtained a PhD in chemical physics from the University of Arizona as a US National Science Foundation graduate research fellow, May 2015, and a Bachelor of Science in both mathematics and chemistry, May 2010. She remains a Designated Scientific Research Campus Colleague at the University of Arizona. She has published over 30 manuscripts in top peer-reviewed journals regarding non-Born-Oppenheimer quantum mechanical finite-nuclear mass variational algorithms and has been cited over 400 times; H-index and i10-index of 10.

Alain Chancé is business advisor to ODE, L3C and is the founder and CEO of Quantalain SASU, a management consulting startup. He has over 30 years of experience in major enterprise transformation projects with a focus on data management and governance gained in major management consulting firms. He has a diploma ingénieur civil des Mines from École des Mines de Saint-Étienne (1981).

He is a Qiskit® Advocate and is an IBM Certified Associate Developer - Quantum Computation using Qiskit® v0.2X since 2021. He has completed a number of hackathons pertaining to quantum computing since 2018.

Acknowledgments

*We would like to thank the technical reviewer, Bruno Fedrici, PhD,
Professor Ludwik Adamowicz for insights into quantum education,
Robert and Suzanne Scifo for artistic renderings of physics and chemistry
concepts as it relates to historical quotes in the chapters, Mellissa Larson
for photographing the lead author, and Quantum Interns of ODE, L3C,
who proofread the chapters and assisted in compiling the glossary: Bhagya
Gopakumar and Sneha Thomas.*

About the reviewer

Bruno Fedrici has a PhD in quantum engineering from the University of Nice Sophia Antipolis along with a university certificate in digital transformation from the University of Lyon. He contributes to the public and business awareness of quantum technologies by providing a bridge between higher education, research, and industry. For three years now, he has been introducing quantum computing basics and quantum-safe security solutions to executives and technical leaders as well as to computer science and engineering students.

Bruno is a lecturer in quantum information science at INSA Lyon. He has also launched Quantum for Everyone, a new online course for non-technical business professionals. Currently Bruno is also a program manager at Quantum Business Europe, a new event focusing on end user applications of quantum technologies.

Table of Contents

3

Quantum Circuit Model of Computation

4

Molecular Hamiltonians

5

Variational Quantum Eigensolver (VQE) Algorithm

6

Beyond Born-Oppenheimer

7

Conclusion

8

References

9

Glossary

Appendix A

Readying mathematical concepts

Appendix B

Leveraging Jupyter Notebooks on the Cloud

Preface

"Learning is finding out what you already know. Doing is demonstrating that you know it. Teaching is reminding others that they know just as well as you. You are all learners, doers, teachers."

– *Richard Bach*

Figure 0.1 – Learning quantum computing and quantum chemistry [authors]

This book aims to demystify quantum chemistry and computing, discuss the future of quantum technologies based on current limitations, demonstrate the usefulness and shortcomings of the current implementations of quantum theory, and share our love of the topic.

This book is not a traditional presentation of quantum chemistry nor quantum computing, but rather an explanation of how the two topics intertwine through the illustration of the postulates of quantum mechanics, particularly with Python code and open-source quantum chemistry packages.

Quantum chemistry has many applications in industry, from pharmaceutical design to energy creation and the development of quantum computing in recent years. With adequate knowledge of quantum chemistry and the postulates of quantum mechanics, we can overcome some of the major hurdles humanity faces and achieve positive impacts. We hope that you can learn sufficient details to be a part of the new and productive solutions moving forward.

Readers we target

All kinds of readers are welcome. However, the people who will benefit the most are those interested in chemistry and computer science at the early stages of learning; advanced high school and early college students, or professionals wanting to acquire a background in quantum chemistry as it relates to computing, both from an algorithm and hardware standpoint. We also summarize useful mathematics and calculus as it relates to solving chemistry problems. The topics will appeal to people of various industry verticals who are interested in a career in quantum computational chemistry and computing.

You will be at the forefront of exciting state-of-the-art opportunities to expand your ideas and start experimenting with your simulations.

A fast path to using quantum chemistry

We chose to write this book in such a way as to demystify the fundamentals of quantum concepts for a curious audience. This book introduces the basics of quantum chemical concepts by describing the five postulates of quantum mechanics, including how these concepts relate to quantum information theory, including basic programming examples of atomic and molecular systems with Python, SimPy [Simpy], QuTiP [QuTiP], and open-source quantum chemistry packages PySCF [PySCF], ASE [ASE_0], PyQMC [PyQMC], Psi4 [Psi4_0], and Qiskit [Qiskit] code. An introductory level of understanding Python is sufficient to read the code, and a browser is all that is required to access the Google Colaboratory and run the companion Jupyter notebooks we provide in the cloud. Each chapter includes an artistic rendering of quantum concepts related to historical quotes.

Through the 1990s, 2000s, and 2010s, there has been amazing progress in the development of computational chemistry packages and, most recently, Qiskit Nature [Qiskit_Nature] [Qiskit_Nat_0]. We outline and introduce basic quantum chemical concepts that are discussed in a modern fashion and relate these concepts to quantum information theory and computation. We use Python, PySCF, and Qiskit Nature for illustrative purposes.

Quantum chemistry

The fundamentals of quantum mechanics and the five postulates directly impact material research and computational chemistry for finding new drugs and catalysts, enabling efficient and cleaner processes for converting chemicals from one form to another. Quantum chemistry is also essential for designing future quantum computers that use the properties of atoms and/or ions. However, quantum chemistry remains an elusive topic that seemingly takes many years to master.

We think that the traditionally long-term achievement of literacy of the topic is directly related to the perceived complexity of the topic and historical approximations made to increase accessibility and usability with conventional computing. With approximation in place and wide acceptance by the scientific community as the only way forward, some fundamental concepts are often overlooked, misunderstood, and eliminated from the disciplines depending on these ideas. We see this as an opportunity to share our love of quantum chemistry in its full potential to enhance the friendliness of and approachability of the topic.

We will share sufficient details so that you understand the limitations that were historically established. For instance, we present a general formulation of the Pauli exclusion principle for all elementary particles that also holds for composite particles, which many textbooks do not adequately explain.

There is more to the quantum story but too much to be included as a first book for the curious. Therefore, we plan to write a following book that expands cutting-edge quantum ideas that are not yet widely used in the scientific community.

How to navigate the book

We advise you to follow the sequential ordering of chapters and gradually master the concepts, methods, and tools that will be useful later in the book.

- *Chapter 1, Introducing Quantum Concepts*, presents a history of quantum chemistry and quantum computing, and introduces the fundamental building blocks of nature, particles and matter, light and energy, and quantum numbers.

- *Chapter 2, Postulates of Quantum Mechanics*, gives a non-expert in quantum physics the concepts, definitions, and notation of quantum mechanics and quantum information theory necessary to grasp the content of this book.

- *Chapter 3, Quantum Circuit Model of Computation*, introduces the quantum circuit model of computation and Qiskit Nature, an open-source framework that provides tools for computing ground state energy, excited states, and dipole moments of molecules.

- *Chapter 4, Molecular Hamiltonians*, presents the molecular Hamiltonian, modeling the electronic structure of a molecule and fermions to qubit mappings.

- *Chapter 5, Variational Quantum Eigensolver (VQE) Algorithm*, shows a process for solving the ground state of a molecule, focusing on the Hydrogen molecule, illustrated with the Variational Quantum Eigensolver (VQE) algorithm using Qiskit Nature.

- *Chapter 6, Beyond Born-Oppenheimer*, gives a glimpse of the beyond Born-Oppenheimer approaches that have not yet been popularized.

- *Chapter 7, Conclusion*, is the opening to the next book.

- *Chapter 8, References*, provides a consolidated list of all the references given at the end of each chapter.

- *Chapter 9, Glossary*, provides a convenient way to look up terms.

- *Appendix A, Readying Mathematical Concepts*, introduces concepts with illustrations in Python code.

- *Appendix B, Leveraging Jupyter Notebooks in the Cloud*, explains how to use free environments on the cloud to run the companion Jupyter notebooks we provide.

- *Appendix C, Trademarks*, lists all the trademarks of the products used in this book.

To get the most out of this book

With the following software and hardware list you can access the Google Colaboratory (Colab), which is a free Jupyter Notebook environment that runs entirely in the cloud and provides online, shared instances of Jupyter notebooks without having to download or install any software:

Chapter	Software required	OS required
0 to 9, Appendix A, B, C	Latest browser	Windows, Mac OS X, or Linux (Any)

Download the example code files

You can download the example code files for this book from GitHub at `https://github.com/PacktPublishing/Quantum-Chemistry-and-Computing-for-the-Curious`. If there's an update to the code, it will be updated in the GitHub repository.

To download the full version of the companion notebooks you can scan the following QR code or go to the provided link to download them.

`https://account.packtpub.com/getfile/9781803243900/code`

We also have other code bundles from our rich catalog of books and videos available at `https://github.com/PacktPublishing/`. Check them out!

Conventions used

There are a number of text conventions used throughout this book.

`Code in text`: Indicates code words in text, database table names, folder names, filenames, file extensions, pathnames, dummy URLs, user input, and Twitter handles. Here is an example: "There is no loop in a quantum circuit, but we can have a classical loop that appends a quantum sub-circuit. In Qiskit we use the `QuantumRegister` class to create a register of qubits and the `QuantumCircuit` class to create a quantum circuit."

A block of code is set as follows:

```
q = QuantumRegister(2)
qc = QuantumCircuit(q)

qc.h(q[0])
qc.cx(q[0], q[1])

qc.draw(output='mpl')
```

Any command-line input or output is written as follows:

```
Mo: 1s² 2s² 2p⁶ 3s² 3p⁶ 4s² 3d¹⁰ 4p⁶ 5s² 4d⁴
```

Get in touch

Feedback from our readers is always welcome.

General feedback: If you have questions about any aspect of this book, email us at `customercare@packtpub.com` and mention the book title in the subject of your message.

Errata: Although we have taken every care to ensure the accuracy of our content, mistakes do happen. If you have found a mistake in this book, we would be grateful if you would report this to us. Please visit `www.packtpub.com/support/errata` and fill in the form.

Piracy: If you come across any illegal copies of our works in any form on the internet, we would be grateful if you would provide us with the location address or website name. Please contact us at `copyright@packt.com` with a link to the material.

If you are interested in becoming an author: If there is a topic that you have expertise in and you are interested in either writing or contributing to a book, please visit `authors.packtpub.com`.

References

[ASE_0] Atomic Simulation Environment (ASE), https://wiki.fysik.dtu.dk/ase/index.html

[NumPy] NumPy: the absolute basics for beginners, https://numpy.org/doc/stable/user/absolute_beginners.html

[Psi4_0] Psi4 manual master index, https://psicode.org/psi4manual/master/index.html

[PyQMC] PyQMC, a python module that implements real-space quantum Monte Carlo techniques, https://github.com/WagnerGroup/pyqmc

[PySCF] The Python-based Simulations of Chemistry Framework (PySCF), https://pyscf.org/

[Qiskit] Qiskit, https://qiskit.org/

[Qiskit_Nat_0] Qiskit_Nature, https://github.com/Qiskit/qiskit-nature/blob/main/README.md

[Qiskit_Nature] Introducing Qiskit Nature, Qiskit, Medium, April 6, 2021, https://medium.com/qiskit/introducing-qiskit-nature-cb9e588bb004

[QuTiP] QuTiP, Plotting on the Bloch Sphere, https://qutip.org/docs/latest/guide/guide-bloch.html

[Simpy] SimPy Discrete event simulation for Python, https://simpy.readthedocs.io/en/latest

Share Your Thoughts

Once you've read *Quantum Chemistry and Computing for the Curious*, we'd love to hear your thoughts! Scan the QR code below to go straight to the Amazon review page for this book and share your feedback.

https://packt.link/r/1-803-24390-2

Your review is important to us and the tech community and will help us make sure we're delivering excellent quality content.

1

Introducing Quantum Concepts

"There are children playing in the streets who could solve some of my top problems in physics, because they have modes of sensory perception that I lost long ago."

– Robert J. Oppenheimer

Figure 1.1 – Girl looking at the image of an atom [Adapted from image licensed by Getty]

Predicting the behavior of matter, materials, and substances not yet measured experimentally is an exciting prospect. Modern computational tools enable you to conduct virtual experiments on freely available resources. Understanding modern models of how chemistry works is essential if you want to get results that match the way Nature operates.

Classical physics works fine for predicting the trajectory of a ball with Newton's law of gravitation or the trajectory of planets around the sun. However, a more accurate description of Nature, especially chemistry, can be found via quantum physics, which encapsulates the postulates of **quantum mechanics**, the foundation of quantum chemistry, and quantum computing. To gain the next level of understanding of chemistry predictions, quantum chemistry algorithms need to be designed to achieve a high level of accuracy. It is not enough to simply program approximate methods and have a quantum computer run them to achieve higher accuracy than the same method implemented on a classical computer.

The postulates of quantum physics are not considered laws of nature and cannot be proven either mathematically or experimentally; rather, they are simply guidelines for the behavior of particles and matter. Even though it took a few decades for these postulates to be formulated and a century for them to be understood by the broader scientific community, they remain a powerful tool for predicting the properties of matter and particles and are the foundation of quantum chemistry and computing.

This chapter is not an exhaustive presentation of the entire history of quantum physics; however, we will mention some of the crucial figures and introduce the topics that we think are the most influential in the 20th century. We discuss the fundamental concepts of particles and the composition of matter, the physical properties of light and its behavior, plus energy and its relation to matter. We extend these concepts to present the quantum numbers related to certain types of chemical applications and properties that can be specifically used for the advancement of quantum computing and predicting states of matter.

In this chapter, we will cover the following topics:

- *Section 1.1, Understanding the history of quantum chemistry and mechanics*
- *Section 1.2, Particles and matter*
- *Section 1.3, Quantum numbers and quantization of matter*
- *Section 1.4, Light and energy*
- *Section 1.5, A brief history of quantum computation*
- *Section 1.6, Complexity theory insights*

Technical requirements

A companion Jupyter notebook for this chapter can be downloaded from GitHub at `https://github.com/PacktPublishing/Quantum-Chemistry-and-Computing-for-the-Curious`, which has been tested in the Google Colab environment, which is free and runs entirely in the cloud, and in the IBM Quantum Lab environment. Please refer to *Appendix B – Leveraging Jupyter Notebooks in the Cloud*, for more information.

1.1. Understanding the history of quantum chemistry and mechanics

Knowing the development of quantum chemistry during the early part of the 20th century is important in order to understand how the postulates of quantum mechanics were discovered. It will also help you grasp the major approximations that have enabled us to achieve scientific milestones. We will mention concepts that will be discussed and described in later chapters of the book, so don't be concerned if you don't understand what the ideas mean or imply. We want to simply start using the terminology of quantum concepts to give some context to the five postulates of quantum mechanics presented in the rest of the book.

Figure 1.2 – Robert J. Oppenheimer – Ed Westcott (U.S. Government photographer), Public domain, via Wikimedia Commons

Quantum mechanics has been a disruptive topic of conversation in the scientific community for just over a century. The most notable controversy of quantum mechanics is that it gave rise to the atomic bomb during World War II. Robert J. Oppenheimer (*Figure 1.2*), considered the father of the atomic bomb, is also the inventor of one of the most widely used and influential approximation to date: the **Born-Oppenheimer (BO) approximation** of 1926 [Intro_BOA_1] [Intro_BOA_2]. This will be described in depth in *Chapter 6, Beyond Born-Oppenheimer*. The BO approximation assumes that the motions of the nuclei are uncoupled from the motions of the **electrons** and led to the formulation of the majority of the computational techniques and software packages available to date, including the basic design of a **qubit** used for quantum computing.

By the time Oppenheimer published his PhD thesis on the BO approximation with Max Born, his academic advisor, many scientists had contributed to **quantum chemistry**. The term *quantum mechanics* appeared for the first time in Born's 1924 paper *Zur Quantenmechanik* [Born]. Quantum mechanics was formulated between 1925 and 1926 with other major contributions from the following:

1. Max Planck for the Planck constant and the Planck relation (*Section 1.4, Light and energy*)

2. Louis de Broglie for the de Broglie wavelength (*Section 1.3, Quantum numbers and quantization of matter*)

3. Werner Heisenberg for the Heisenberg uncertainty principle (*Section 1.4, Light and energy*)

4. Erwin Schrödinger for the Schrödinger equation (*Section 1.4, Light and energy*)

5. Paul Dirac for the Dirac equation, a relativistic wave equation for fermionic systems, and for the Dirac notation, also known as bra-ket notation (*Section 1.3, Quantum numbers and quantization of matter*)

6. Wolfgang Pauli for the Pauli exclusion principle (*Section 1.3, Quantum numbers and quantization of matter*)

These scientists attended the 5th Solvay conference on quantum mechanics (*Figure 1.3*) along with other very influential scientists that are not discussed. This image captures the first cohort of quantum scientists that had a great influence on the 20th century.

Figure 1.3 – Solvay Conference on quantum mechanics, 1927. Image is in the public domain

The BO approximation was a necessary development primarily because of the **Pauli exclusion principle** (**PEP**), which was formulated in 1925. Pauli described the PEP for electrons, and it states that it is impossible for two electrons of the same **atom** to simultaneously have the same values of the following four quantum numbers: n, the **principal quantum number**; l, the **angular momentum quantum number**; m_l, the **magnetic quantum number**; and m_s, the **spin quantum number**. His work has been further extended to bosonic particles. The PEP leads to a certain type of computational complexity that initiated the necessity for the BO approximation; see *Section 1.6, Complexity theory insights* for more details. We will go into the detail of quantum quantities and describe PEP for different particle types in *Section 1.3, Quantum numbers and quantization of matter*.

The rapid development by the aforementioned group of thought leaders came about with the important groundwork laid out by their predecessors and their discoveries related to the hydrogen atom – the simplest of all elements of the periodic table:

- Johan Balmer in 1885 discovered the Balmer emission line series [Balmer_series].

- Johannes Rydberg in 1888 generalized the Balmer equation for all transitions of hydrogen [Chem_spectr].

- Theodore Lyman from 1906 to 1914 discovered the Lyman series of hydrogen atom spectral lines in the ultraviolet [Lyman_series].
- Friedrich Paschen in 1908 discovered the Paschen spectral lines in the infrared band [Chem_spectr].

The structure of the hydrogen atom will be discussed in *Section 1.4, Light and energy,* and outlined computationally in *Chapter 5, Variational Quantum Eigensolver (VQE) Algorithm.*

The work of Johannes Rydberg led to the definition of the fundamental constant used in spectroscopy. Rydberg worked side-by-side with Walter Ritz in 1908 to develop the Rydberg-Ritz combination principle about the relationship between frequencies and the spectral lines of elements [Rydberg-Ritz]. Rydberg states of atoms are used in quantum computation, and this is discussed in *Chapter 3, Quantum Circuit Model of Computation.*

A year after the development of the Rydberg-Ritz combination principle, Ritz developed a way to solve the **eigenvalue** problem [Rayleigh–Ritz], which is widely used today in the field of computational chemistry and is known as the **Rayleigh-Ritz variational theorem**. This method is the inspiration for the **Variational Quantum Eigensolver (VQE)** discussed in detail in *Chapter 5, Variational Quantum Eigensolver (VQE) Algorithm.*

In conjunction with the testing of the Rayleigh-Ritz variational method mechanically by John William Strutt, 3rd Baron Rayleigh, known for the Rayleigh scattering of light, this method is called the Rayleigh-Ritz method even though it was written and formulated by Ritz. In short, it allows for the approximation of the solutions to the eigenvalue problem. His work led to the method of applying the superposition principle to approximate the **total wave function**; this mathematical expansion is one of the postulates of quantum mechanics described in *Chapter 2, Postulates of Quantum Mechanics.*

With a better understanding of the hydrogen atom, in 1913, Niels Bohr attempted to describe the structure of atoms in more detail with fundamental concepts about quantization and quantum theory [Bohr_1] [Bohr_2]. He received the Nobel Prize in 1922 for his Bohr model. In his dissertation, many articles verify, predict, and assess very accurate Rydberg states of small atoms ($Z < 7$) as well as the rotational-vibrational (rovibrational) states of small molecules. Bohr's atomic model describes the electron energy transitions starting from the second, first, and third atom electron layers, known as Balmer series, Lyman series, and Paschen series, and the corresponding hydrogen emission spectrums discovered previously.

In the 1930s, Linus Pauling and Edgar Bright Wilson Jr. popularized quantum mechanics as it is currently applied to chemistry [Pauling]. Pauling eventually received a Nobel Prize for Chemistry in 1954, and later on, he received a Nobel Peace Prize in 1964 for his political activism regarding quantum mechanics.

The development of the postulates of quantum mechanics since these major contributions has remained, in general, the same to date.

Thanks to the development of classical computers and clever computational methods, many computational chemistry packages have been produced to further our understanding of chemistry. Some notable methods, other than the **Rayleigh-Ritz variational theorem, a**re the **Quantum Monte Carlo (QMC)** [QMC], **Hartree-Fock (HF) method**, **Coupled-cluster (CC)**, and **Density-functional theory (DFT)**, among others. In this book, we will illustrate some of these methods with Python and open-source quantum chemistry packages such as PySCF, ASE, PyQMC, Psi4, and Qiskit in subsequent chapters.

In the late 20th century, Richard Feynman stated that quantum concepts could be used for quantum computing [Preskill_40y]. Physicists Jonathan Dowling and Gerard Milburn wrote in 2002 that we have entered a second quantum revolution, actively employing quantum mechanics in quantum information, quantum sensing, quantum communication, and analog quantum simulation [Dowling]. We will summarize the history of quantum computation in *Section 1.5, A brief history of quantum computation*. This second quantum revolution is seen as a way to overcome computational complexity using matter and the postulates of quantum mechanics.

The question becomes: what is the purpose of implementing approximate methods in a quantum computer? Are quantum computers supposed to help us in reaching beyond the aforementioned methods? We intend to address these questions in this book, specifically in *Chapter 6, Beyond Born-Oppenheimer*.

1.2. Particles and matter

In general, particles and matter have three unique properties that do not change: mass, charge, and magnetic spin. For some particles, these properties can have a value of zero; otherwise, these properties are real numbers and can be measured experimentally. Mass can only be positive, while charge can be positive or negative.

In the following subsections, we will review elementary and composite particles, which include both fermions and bosons. Understanding these kinds of particles is fundamental to the understanding of quantum chemistry and the potential use of quantum computing.

Elementary particles

Elementary particles are either fermions or bosons [Part_1]. The term *fermion* was coined by Dirac, who was inspired by the physicist Enrico Fermi. Elementary boson particles are part of the Standard Model [Std_model] and do not necessarily take part in quantum chemistry, but rather fundamental physics.

The electron (e^-) is the primary elementary fermionic particle associated with quantum chemistry. Electrons have a mass of 9.1093837015 x 10^{-31} kilograms (kg) [e_mass] and an electric charge of negative one (-1). The size of the electron is on the order of approximately 10-15 centimeters (cm). In most computational methods simulations, we change the reference mass so that an electron's mass is equal to 1, making the computation easier. There are also the muon (μ^-) and tau (τ^-) particles, which have a negative one electric charge (-1) but are much heavier than the electron. The associated antiparticles, the positron (e^+), antimuon (μ^+), and antitau (τ^+), have the same mass but opposite electric charge (+1) to their counterparts. The standard computational model can only handle the electron. Recent advanced scientific programs are capable of handling electronic substitutions with muons and taus and the antiparticles.

The current view of how matter is structured is depicted in *Figure 1.4*, using a hydrogen atom as the simplest example. We depict quarks in this image, but to learn more about this, visit CERN [CERN_quark]. Please note that we have included a fuzzy electron cloud around the **nucleus** and are trying to move away from the old model of an electron following a well-defined trajectory.

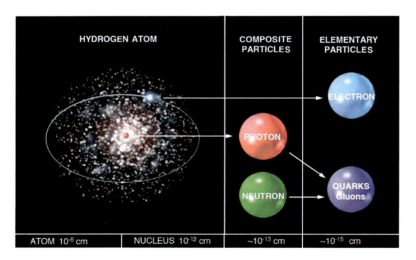

Figure 1.4 – Scaling structure of the hydrogen atom [authors]

Composite particles

The composite particles that contribute mostly to quantum chemistry and computing are atomic nuclei, atoms, and molecules, all of which can be either fermions or bosons. Fermion and boson particles obey the PEP, which is discussed in more detail in *Section 1.3, Quantum numbers and quantization of matter.*

Atomic nuclei

The building blocks of atomic nuclei are **nucleons**. Nucleons are **protons** and **neutrons**. A nucleus contains one or more protons and zero or more neutrons, which are held together by the strong nuclear force. The size of protons and neutrons is in the order of $\sim 10^{-13}$ cm, whereas atomic nuclei range from $\sim 10^{-13}$ cm to about $\sim 10^{-12}$ cm.

Protons have an electric charge of positive one (+1), which is equal in magnitude to that of an electron and have a mass that is 1,836.15267343 times greater than the electron [proton-electron-mass-ratio].

Neutrons have an electric charge of zero and a mass that is 1,838.68366173 times greater than the electron [Neutron-electron-mass-ratio]. The neutron is slightly heavier than the proton.

The number of protons in the nuclei determines the type of element in the periodic table (*Figure 1.6*). The hydrogen atom is the only element that does not contain a neutron in the nucleus. Isotopes of the elements are determined by varying the number of neutrons within the nucleus. The isotopes for hydrogen are deuterium and tritium. Isotopes play an important role in quantum chemistry as the quantum properties can vary. This is an important aspect in computational chemistry as nuclear effects have an impact using the BO approximation (covered in more detail in *Chapter 6, Beyond Born-Oppenheimer*).

Atoms

An atom defines the chemical properties of the bulk matter. Atoms are the combination of atomic nuclei and the electrons that move about outside the nuclei. An atom has no overall electric charge since the number of electrons is equal to the number of protons. The size of atoms is of the order of $\sim 10^{-8}$ cm. An ion is an atom with a net electric charge, either positive or negative, acquired by losing or gaining one or more electrons. Atomic isotopes can also lose or gain electrons and be considered ions. If an atom has gained an electron, the atom will have a negative charge; conversely, it will become positively charged whenit loses an electron. The size of ions can vary. A positive ion is called a cation, and a negative ion is called an anion. Atoms, isotopes, and ions are core topics of quantum chemistry and computing.

Molecules

A molecule is the smallest unit of matter that maintains the chemical properties of a substance. Molecules are composed of two or more atoms and/or isotopes of atoms. They are considered the smallest building blocks of substances that maintain the chemical properties of more than one molecule of the substance. Molecules can also be ions as they can also lose and gain electrons. A molecule is one of the most basic units of matter.

1.3. Quantum numbers and quantization of matter

Quantization is the concept that matter, particles, and other physical quantities, such as charge, energy, and magnetic spin, can have only certain countable values. These certain countable values can be either discrete or continuous variables. Discrete values are defined as countable in a *finite* amount of time. Continuous values are defined as countable in an *infinite* amount of time. Whether or not a quantum system is discrete or continuous depends on the physical system or the observable quantity.

We will discuss the particles that are most associated with quantum chemistry: protons, neutrons, electrons, and hydrogen atoms. The neutrons and protons comprise the nucleus of atoms and are held together by the strong nuclear force, and they do not have a measurable angular momentum quantum number within the nucleus. In contrast, free protons and neutrons not bound within a nucleus can be in motion and then possess an angular momentum quantum number. Within a nucleus, all the protons and neutrons couple (or add) together with their given magnetic quantum numbers so that the nucleus has an overall magnetic quantum number. This is also true for the spin momentum (S) quantum number for these particles, which for each is 1/2. In general, we consider the overall magnetic and spin momentum quantum numbers of a nucleus and not the individual protons and neutrons of the nucleus.

Electrons in an atom

The following five quantum numbers correspond to electrons in an atom:

- The principal quantum number, n, describes the energy level or the electron's position in a shell of the atom and is numbered from 1 up to the shell containing the outermost electron of that atom. Technically, n can range from 1 to infinity, thus it is a continuous quantum number. However, as the electron is excited to higher and higher values of n, and dissociates from the atom, it is then considered a free electron, and an ion. This process is called ionization, and n is then considered discrete.

- The angular momentum quantum number, l, also known as the orbital quantum number or the azimuthal quantum number, describes the electron subshell and gives the magnitude of the orbital angular momentum through the relation: $L^2 = \hbar^2 l(l + 1)$. In chemistry and spectroscopy, $l = 0$ is called the s orbital, $l = 1$ the p orbital, $l = 2$ the d orbital, and $l = 3$ the f orbital. Technically, there are more orbitals beyond the f orbital, that is, $l = 4 = g$, $l = 5 = h$, and so on, and are of higher energy levels.

- The magnetic quantum number, m_l, describes the electron's energy level within its subshell and the orientation of the electron's orbital. It can take on integer values ranging from $-l, ..., 0, ..., +l$.

- The spin quantum number, s, varies for each particle type, and there is no classical analog to describe what it is. The spin quantum number describes the intrinsic spin momentum of a certain particle type; for the electron, it is equal to 1/2.

- The **spin projection quantum number**, m_s, gives the projection of the spin momentum s along the specified axis as either "spin up" (+½) or "spin down" (-½) in a given spatial direction. This direction in quantum computing is defined as the z-axis.

The wave function and the PEP

A wave function is a mathematical tool that describes the state, or the motion, and the physical properties of particles and matter. This concept is the first postulate of quantum mechanics. We will go into the details of this in *Chapter 2, Postulates of Quantum Mechanics*. The variables of the wave function are the quantum numbers previously described, as well as position and time. The PEP ensures that the full wave function for a given system is complete and is different for fermions, bosons, and composite particles.

Fermions

In 1925, Wolfgang Pauli stated that in a single atom, no two electrons can have an identical set of quantum numbers, n, l, m_l, and m_s. This principle was supplemented by stating that only antisymmetric pair permutations of the electronic wave function are allowed [Kaplan] [Dirac_2]. Antisymmetric refers to obtaining a minus sign (-) of the wave function upon applying the permutation operator to the wave function.

Bosons

The only possible states of a system of identical bosons are those for which the total wave function is symmetric [Kaplan]. Symmetric refers to obtaining a plus sign (+) of the wave function upon applying the permutation operator wave function.

Composite particles

The following general formulation of the PEP for all elementary particles also holds for composite particles [Kaplan]: *The only possible states of a system of identical particles possessing spin S are those for which the total wave function is symmetric for integer values of S (the Bose–Einstein statistics) and antisymmetric for half-integer values of S (the Fermi–Dirac statistics).*

Dirac notation

Dirac notation is also known as bra-ket notation. The state of a quantum system, or the wave function, is represented by a ket, $|x\rangle$, which is a column vector of coordinates and/or variables. The bra, $\langle f|$, denotes a linear function that maps each column vector to a complex conjugate row vector. The action of the row vector $\langle f|$ on a column vector $|x\rangle$ is written as $\langle f|x\rangle$. In *Chapter 2, Postulates of Quantum Mechanics*, we will show how Dirac notation relates to the aforementioned quantum numbers. Dirac notation will be further explained and illustrated in *Chapter 3, Quantum Circuit Model of Computation*.

1.4. Light and energy

Light and energy are fundamental to the behavior of matter. In this section, we will outline how light and energy are related to mass, momentum, velocity, wavelength, and frequency. We will also introduce electronic transitions of the hydrogen atom.

Planck constant and relation

In 1900, German physicist Max Planck explained the spectral-energy distribution of radiation emitted by a black body by assuming that the radiant energy exists only in discrete quanta that are proportional to the frequency. The Planck relation states that the energy (E) of a photon is proportional to its frequency (ν) and inversely proportional to the wavelength (λ): $E = h\nu = hc/\lambda$. h is the Planck constant, h =6.62607015×10^{-34} Joule x Hertz^{-1} (J x Hz^{-1}), with Hertz being defined as inverse seconds (Hz = s^{-1}), and c is the speed of light, which is equal to 299,792,458 meters per second (ms^{-1}).

The de Broglie wavelength

The de Broglie wavelength formula relates the mass (m), momentum (P), and velocity (v) of a particle to the wavelength (λ): $\lambda = \dfrac{h}{p} = \dfrac{h}{mv}$, and it is the scale at which a particle behaves like a wave.

Heisenberg uncertainty principle

The uncertainty principle is related to the associated accuracy of measuring physical quantities. We define the accuracy of measuring physical quantities using the standard deviation (σ), or average variance of a given set of measurements. The uncertainty principle asserts a lower bound on the accuracy for predicting certain pairs of physical quantities of a particle from initial conditions. The more accurately you know one quantity, the less you know about the other quantity.

For instance, momentum (P) and position (x) are a pair of physical quantities that follow uncertainty such that if you know exactly where a particle is, the less you know about its momentum. Conversely, the more you know about its momentum, the less you know about exactly where it is.

The standard deviation of position (σ_x) and the standard deviation of momentum (σ_p) are related by the following inequality, $\sigma_x \sigma_p \geq \dfrac{\hbar}{2}$, where $\hbar = \dfrac{h}{2\pi}$ is the reduced Planck constant.

Energy levels of atoms and molecules

Transitions between different electron energy levels in an atom, or between different vibration or rotation energy levels in a molecule, occur by the processes of absorption, emission, and/or stimulated emission of photons. Only when the energy of a photon matches the exact difference in the energy between the initial and final states does a transition occur. In an atom, the energy associated with an electron is that of its orbital. High energy states close to when an atom is being ionized are called Rydberg states. An atomic or molecular orbital describes the probability of finding an electron at a given point in space in an atom or molecule. In the simplest atom, when a hydrogen atom absorbs a photon of light, the electron jumps to a higher energy level, for example from $n = 1$ to $n = 2$. Conversely, a photon is emitted when an electron jumps to a lower energy level, for instance from $n = 3$ to $n = 2$ [Byjus].

Hydrogen spectrum

The hydrogen spectrum has been divided into spectral lines [Chem_spectr]:

- The Lyman series corresponds to transitions from excited states $n > 1$ to $n = 1$.

- The Balmer series corresponds to transitions from excited states $n > 2$ to $n = 2$.

- The Paschen series corresponds to transitions from excited states $n > 3$ to $n = 3$.

- The Brackett series corresponds to transitions from excited states $n > 4$ to $n = 4$.

- The Pfund series corresponds to transitions from excited states $n > 5$ to $n = 5$.

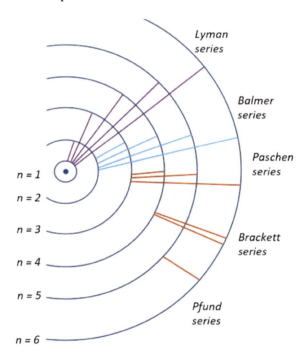

Figure 1.5 – Emission spectrum of hydrogen

Rydberg constant and formula

The Rydberg formula, also known as the Rydberg-Ritz recombination principle, calculates the inverse of the wavelengths (λ) of element spectral lines [Chem_spectr]:

$$\frac{1}{\lambda} = RZ^2 \left(\frac{1}{n_1^2} - \frac{1}{n_2^2} \right)$$

where R is the Rydberg constant, Z is the **atomic number**, n_1 is the principal quantum number of the lower energy level, and n_2 is the principal quantum number of the upper energy level. The Rydberg constant for heavy atoms is $R = 10{,}973{,}731.568160(21)$ meters^{-1} (m^{-1}) [Rydberg_R].

Electron configuration

In the early 1920s, Niels Bohr and Wolfgang Pauli formulated the Aufbau principle (from German, Aufbauprinzip, *building-up principle*), which states that electrons fill subshells of the lowest energy available before filling subshells of higher energy. The Aufbau principle is based on the Madelung rule, which states that electrons fill the orbitals in order of increasing $n + l$ such that whenever two orbitals have the same value of $n + l$, they are filled in order of increasing n. The symbols used for writing the electron configuration start with the energy level n followed by the **atomic orbital** letter, and finally the superscript, which indicates how many electrons are in the orbital. For example, the notation for phosphorus (P) is $1s^2\, 2s^2\, 2p^6\, 3s^2\, 3p^3$.

Calculating the electron configuration of atomic elements using the Madelung rule

The following Python program calculates the electron configuration of all elements up to Rutherfordium (N=104) using the Madelung rule. It is derived from a program published by Christian Hill on his website, Learning Scientific Programming with Python, Question P2.5.12 [Hill].

Setting up a list of atomic symbols

The following array contains a list of atomic symbols:

```
atom_list = ['H', 'He', 'Li', 'Be', 'B', 'C', 'N', 'O', 'F',
'Ne', 'Na','Mg', 'Al', 'Si', 'P', 'S', 'Cl', 'Ar', 'K', 'Ca',
'Sc', 'Ti', 'V', 'Cr', 'Mn','Fe', 'Co', 'Ni', 'Cu', 'Zn', 'Ga',
'Ge', 'As', 'Se', 'Br', 'Kr', 'Rb', 'Sr','Y', 'Zr', 'Nb', 'Mo',
'Tc', 'Ru', 'Rh', 'Pd', 'Ag', 'Cd', 'In', 'Sn', 'Sb','Te', 'I',
'Xe', 'Cs', 'Ba', 'La', 'Ce', 'Pr', 'Nd', 'Pm', 'Sm', 'Eu',
'Gd','Tb', 'Dy', 'Ho', 'Er', 'Tm', 'Yb', 'Lu', 'Hf', 'Ta',
'W', 'Re', 'Os', 'Ir','Pt', 'Au', 'Hg', 'Tl', 'Pb', 'Bi', 'Po',
'At', 'Rn', 'Fr', 'Ra', 'Ac', 'Th','Pa', 'U', 'Np', 'Pu', 'Am',
'Cm', 'Bk', 'Cf', 'Es', 'Fm', 'Md', 'No','Lr','Rf']
```

Setting up a list of atomic orbital letters

The following block of code initializes the list of orbitals as introduced in *Section 1.3, Quantum numbers and quantization of matter*:

```
l_orbital = ['s', 'p', 'd', 'f', 'g']
```

Setting up a list of tuples in the order in which the corresponding orbitals are filled

The following block of code initializes `nl_pairs` as a list of tuples that is used to calculate the electron configuration of all atomic elements:

```
nl_pairs = []
for n in range(1,8):
    for l in range(n):
        nl_pairs.append((n+l, n, l))
nl_pairs.sort()
print(nl_pairs[:9])
print(nl_pairs[9:18])
print(nl_pairs[18:len(nl_pairs)])
```

Here's the result:

```
[(1, 1, 0), (2, 2, 0), (3, 2, 1), (3, 3, 0), (4, 3, 1), (4, 4, 0), (5, 3, 2), (5, 4, 1), (5, 5, 0)]
[(6, 4, 2), (6, 5, 1), (6, 6, 0), (7, 4, 3), (7, 5, 2), (7, 6, 1), (7, 7, 0), (8, 5, 3), (8, 6, 2)]
[(8, 7, 1), (9, 5, 4), (9, 6, 3), (9, 7, 2), (10, 6, 4), (10, 7, 3), (11, 6, 5), (11, 7, 4), (12, 7, 5), (13, 7, 6)]
```

Figure 1.6 – List of tuples in the order in which the corresponding orbitals are filled

Initializing a list of orbitals and the electrons they contain with the 1s orbital

The following block of code initializes these variables:

- nl_index: Index of the subshell in the nl_pairs list.

- n_elec: Number of electrons currently in this subshell.

- config: Electronic configuration, an array of arrays. In each of these arrays, the first item is a string concatenating the energy level n followed by the atomic orbital letter, and the second item is the number of electrons in the orbital to be displayed in superscript.

- el_config: Dictionary of electron configurations.

```
nl_idx, n_elec = 0, 0
n, l = 1, 0
config = [['1s', 0]]
el_config = {}
```

The superscript function returns an integer in superscript from 0 to 9:

```
def superscript(n):
    return "".join(["⁰¹²³⁴⁵⁶⁷⁸⁹"[ord(c)-ord('0')] for c in
str(n)])
```

This code calculates the electronic configurations of all atomic elements:

```
for element in atom_list:
    n_elec += 1
    if n_elec > 2*(2*l+1):
        # This subshell is full: start a new subshell
        nl_idx += 1
        _, n, l = nl_pairs[nl_idx]
        config.append(['{}{}'.format(n, l_orbital[l]), 1])
        n_elec = 1
```

```
    else:
        # Add an electron to the current subshell
        config[-1][1] += 1
    # Building configuration string from a list of orbitals and
n_elec
    el_config[element] = ' '.join(['{:2s}{:1s}'.format(e[0],
                        superscript(e[1])) for e in
config])
```

This code prints the electronic configurations of the first five atomic elements:

```
for element in atom_list[:5]:
  print('{:2s}: {}'.format(element, el_config[element]))
```

Here's the result:

```
H: 1s¹
He: 1s²
Li: 1s² 2s¹
Be: 1s² 2s²
B: 1s² 2s² 2p¹
```

Figure 1.7 – Electronic configurations of the first five atomic elements

This code prints the electronic configuration of the element molybdenum (Mo):

```
element = 'Mo'
print('{:2s}: {}'.format(element, el_config[element]))
```

Here's the result:

```
Mo: 1s² 2s² 2p⁶ 3s² 3p⁶ 4s² 3d¹⁰ 4p⁶ 5s² 4d⁴
```

Figure 1.8 – Electronic configuration of the element molybdenum

Schrödinger's equation

Schrödinger's equation can be used to describe the time dynamics (evolution) or static (stationary) states of a quantum mechanical system. For time dynamics, Schrödinger's equation is: $i\hbar\dfrac{d}{dt}|\psi\rangle = \hat{H}|\psi\rangle$, where i is the imaginary unit ($i^2 = -1$), $\frac{d}{dt}$ is the time derivative, $\hat{H}$ is the Hamiltonian operator (an observable that accounts for the total energy of that system and is the sum of the kinetic energy and potential energy), and $|\psi\rangle$ is the state vector (or wave function) of the quantum system as a function of time (t). The time-independent Schrödinger equation can be written as follows, i.e. static: $\hat{H}|\psi\rangle = E|\psi\rangle$, where E is the energy eigenvalue, and $|\psi\rangle$ is the state vector of the quantum system not as a function of time.

Probability density plots of the wave functions of the electron in a hydrogen atom

The electron configuration of an atom or a molecule is described by its orbitals based on the quantum numbers described in *Section 1.3, Quantum numbers and quantization of matter*, which can be depicted by probability clouds. *Figure 1.9* shows the probability density plots of the wave functions of the electron in a hydrogen atom at different energy levels for principal quantum numbers up to 4. We will describe in more detail in *Chapter 2, Postulates of Quantum Mechanics*, why these images look the way they do.

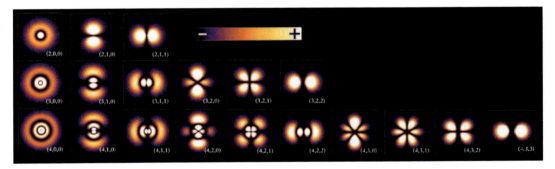

Figure 1.9 – Hydrogen density plots. Each plot indicates (n, l, m_l). Credit: [PoorLeno]. Image is in the public domain

The orbital approximation is a method of visualizing electron orbitals for chemical species that have two or more electrons [Orb_Approx].

Now that we've gone through all the key concepts of quantum chemistry, you have the foundation you need to understand how this relates to quantum computation.

1.5. A brief history of quantum computation

The first revolution was the formulation of the postulates in the early 1900s. Following the first revolution, in 1936, Alan Turing created a theoretical model for automatic machines, now called Turing machines, which laid the theoretical foundations of computer science. In 1980, Paul Benioff published a paper that described a quantum mechanical model of Turing machines [Benioff]. With this and the advancements in quantum chemistry, the foundations were in place for quantum computers.

The first time that quantum computation was discussed within the broader scientific community was when Richard Feynman gave a keynote lecture at a conference called the Physics of Computation held in May 1981 at Massachusetts Institute of Technology (MIT). This keynote lecture discussed harnessing quantum physics to build a quantum computer [Preskill_40y]. In May 2021, on the anniversary of the conference, IBM organized an event called QC40: Physics of Computation Conference 40th Anniversary [QC40], celebrating the second quantum revolution.

A few decades before the first conference, in 1964 John Stewart Bell published a paper "On the Einstein Podolsky Rosen Paradox." He proved that there are no local hidden variables of quantum mechanics [Bell_1], an important development that paved the way for quantum information theory, with applications in quantum computing and quantum communication. This is the basis of using quantum advantage.

Building on these ideas for quantum computation, in 1985, David Deutsch published a paper that laid the foundations of the quantum theory of computation. He stated the Church–Turing–Deutsch principle, which is that a universal computing device can simulate every physical process. He also published the Deutsch–Jozsa algorithm with Richard Jozsa in 1992, which is the first example of a quantum algorithm that is exponentially faster than any possible deterministic classical algorithm [Deutsch-Jozsa].

Peter Shor created a polynomial-time quantum computer algorithm for integer factorization, which is now known as Shor's algorithm [Shor], in 1994. In 1996, Emanuel Knill and Raymond Laflamme developed a general theory of quantum error correction [Knill]. In 2000, David P. DiVincenzo discussed five requirements for the physical implementation of quantum computation and two more requirements pertaining to quantum communication [DiVincenzo]. In 2001, IBM researchers published an experimental realization of Shor's algorithm [Vandersypen].

In 2014, an article by lead author Alberto Peruzzo et al. in Nature Communications introduced the hybrid Variational Quantum Eigensolver (VQE) algorithm for finding an estimate of the ground state energy (lowest energy) of a molecule [VQE_1]. Please be reminded that the variational method was mentioned in *Section 1.1, Understanding thehistory of quantum chemistry and mechanics*. The IBM Quantum team used the VQE algorithm in 2017 to simulate the ground state energy of the lithium hydride molecule [VQE_2]. VQE algorithms are available on several platforms; for instance, Qiskit, enabling experimentation on a variety of quantum processors and simulators, which is used in this book. So far, at the time of this book being published, no claim has been made that a VQE algorithm has outperformed classical supercomputers in computational chemistry based on first principles (ab initio).

On October 23, 2019, John Martinis, Chief Scientist Quantum Hardware, and Sergio Boixo, Chief Scientist Quantum Computing Theory, at Google AI Quantum published the results of a quantum supremacy experiment in the Nature article *Quantum Supremacy Using a Programmable Superconducting Processor* [Arute], which is a controversial claim and suggestion. Quantum supremacy is an experimental demonstration that a quantum computing device can perform a particular computation that no classical computer could do in a reasonable amount of time.

In the current Noisy Intermediate-Scale Quantum (NISQ) era, an ever-increasing number of actors from academia and industry with massive funding programs take part in the so-called race to demonstrate a quantum advantage, either a computational speedup or a reduction in energy costs or both. Current quantum processors have limitations such as too few qubits and limited circuit depths. The goal is to obtain fault-tolerant devices that may need to leverage an advanced understanding of quantum chemistry.

1.6. Complexity theory insights

Complexity theory has two important facets: one is regarding the PEP and the use of the BO approximation, to which we dedicate part of *Chapter 2, Postulates of Quantum Mechanics*; and two is the complexity of computation. This section describes the complexity of computation as it relates to quantum systems.

In his keynote lecture at the Physics of Computation Conference at MIT in 1981 [MIT_QC_1981], Richard Feynman asked the question: *"Can a classical computer simulate either a classical system or a quantum system exactly?"* He also stated that the number of computer elements required to simulate a large physical system should only be proportional to the size of the physical system. Feynman pointed out that calculating the probability of each of R particles of a large quantum system being at each of N points require an amount of memory proportional to N^R, that is, increasing exponentially with R. His next question was whether a classical system could simulate probabilistically a quantum system. Using Bell's theorem, which rules out local hidden variables [Bell_1] [Bell_2], Feynman showed that the answer was again negative.

Feynman suggested that quantum systems could simulate other quantum systems and he challenged the computer scientists to work out the classes of different kinds of quantum mechanical systems that are intersimulatable — that is, equivalent — as has been done in the case of classical computers [Preskill_40y]. Computational complexity theory, a field of theoretical computer science, attempts to classify computational problems according to their resource usage, space, and time.

In this theory, problems that can be solved in polynomial (P) time are in class P, problems for which an answer can be verified in polynomial time are in class nondeterministic polynomial (NP) [PvsNP], and problems for which no efficient solution algorithm has been found and it is easy to check that a solution is correct are in class NP-complete. For example, we do not know an efficient way to decompose a number into its prime components, while it is easy to check that the product of prime numbers is a solution.

The question whether P = NP is in the list of seven Millennium Prize Problems [Clay]. Problems that can be solved by a quantum computer in polynomial time are in class BQP.

In his 2008 article *The Limits of Quantum Computers* [Aaronson_1] [Aaronson_2], Scott Aaronson pointed out that a common mistake is to claim that, in principle, quantum computers could solve NP-complete problems by processing every possible answer simultaneously. We will explain in *Chapter 4, Molecular Hamiltonians*, that a quantum computer can process simultaneously all possible states of a quantum register prepared in superposition. The difficulty is to design algorithms that use quantum interference and entanglement so that the quantum computer only gives the right answers. Scott Aaronson also manages a list of all the complexity classes, Complexity Zoo [Comp_Zoo]. Stephen Jordan manages a quantum algorithm zoo [Qa_Zoo].

A new discovery in complexity theory establishes why the gradient descent algorithm cannot solve some kinds of problems quickly [Fearnley] [Daskalakis]. Training variational quantum algorithms is an NP-hard problem [Bittel]. These results impact all variational quantum algorithms, such as VQE, because they rely on a classical optimizer to optimize the set of parameters of a quantum circuit.

Summary

Using the postulates of quantum mechanics to understand both quantum chemistry and computing will help the next generation of scientists to predict the behavior of matter not yet measured experimentally. This is an exciting prospect for humanity, and we hope that the historical perspective provides a reference point for the current state of the industry revealing new opportunities for progress. There is a circular thought process between quantum chemistry, the use of quantum computing, and the use of both in conjunction.

Questions

Please test your understanding of the concepts presented in this chapter with the corresponding Google Colab notebook.

1. What is the primary elementary fermionic particle associated with quantum chemistry?

2. What value of l (angular momentum quantum number) corresponds to a p orbital?

3. What is the value of the spin quantum number s for an electron?

4. Fermions obey the PEP, which means that the paired particle permutation of the wave function must be antisymmetric. What is the sign for antisymmetry?

5. What is the energy of a photon whose wavelength is 486.1 nanometers?

   In the International System of Units (SI):

   $h =$ 6.62607015×10^{-34} J x Hz^{-1} is the Planck constant

   $c =$ 299,792,458 (ms^{-1}) is the speed of light

   Hint: Look at the blue line in the visible spectrum of the hydrogen atom. You also need to convert from meters to nanometers.

6. To which series of hydrogen atoms does the wavelength in the previous question, 486.1 nanometers, correspond? Lymer, Balmer, or Paschen?

7. Regarding the Rydberg formula, what is the principal quantum number n^2 of the upper energy level corresponding to this transition of 486.1 nanometers?

8. Provide the full electron configuration of the hydrogen element.

   Enter the full electron configuration without superscript.

9. Provide the full electron configuration of the Nitrogen element.

Answers

1. Electron
2. 1
3. ½
4. -
5. 4.086496311764922e-19
6. Balmer
7. 4
8. 1s1
9. 1s2 2s2 2p3

References

[Aaronson_1] Scott Aaronson, The Limits of Quantum Computers, Scientific American, March 2008, `https://www.scientificamerican.com/article/the-limits-of-quantum-computers/`

[Aaronson_2] Scott Aaronson, The Limits of Quantum Computers (DRAFT), `https://www.scottaaronson.com/writings/limitsqc-draft.pdf`

[Arute] Arute, F., Arya, K., Babbush, R. et al., Quantum supremacy using a programmable superconducting processor, Nature 574, 505–510 (2019), `https://doi.org/10.1038/s41586-019-1666-5`

[Balmer_series] Balmer Series, Wikipedia, `https://en.wikipedia.org/wiki/Balmer_series`

[Bell_1] Bell, J. S., On the Einstein Podolsky Rosen Paradox, Physics Physique Fizika 1, 195: 195–200, 1964, `https://doi.org/10.1103/PhysicsPhysiqueFizika.1.195`

[Bell_2] "Chapter 2: On the Einstein-Podolsky-Rosen paradox". Speakable and Unspeakable in Quantum Mechanics: Collected Papers on Quantum Philosophy (Alain Aspect introduction to 1987 ed.), Reprinted in JS Bell (2004), Cambridge University Press. pp. 14–21. ISBN 978-0521523387

[Benioff] Benioff, P., The computer as a physical system: A microscopic quantum mechanical Hamiltonian model of computers as represented by Turing machines, https://doi.org/10.1007/BF01011339

[Bittel] Lennart Bittel and Martin Kliesch, Training variational quantum algorithms is NP-hard — even for logarithmically many qubits and free fermionic systems, DOI:10.1103/PhysRevLett.127.120502, 18 Jan 2021, https://doi.org/10.1103/PhysRevLett.127.120502

[Bohr_1] N. Bohr, I., On the Constitution of Atoms and Molecules, Philosophical Magazine, 26, 1-25 (July 1913), DOI: 10.1080/14786441308634955

[Bohr_2] Bohr's shell model, Britannica, https://www.britannica.com/science/atom/Bohrs-shell-model#ref496660

[Born_1] Born, M., Jordan, P. Zur Quantenmechanik, Z. Physik 34, 858–888 (1925), https://doi.org/10.1007/BF01328531

[Byjus] BYJU'S, Hydrogen spectrum, Wavelength, diagram, Hydrogen emission spectrum, https://byjus.com/chemistry/hydrogen-spectrum/#

[CERN_quark] CERN Voyage into the world of atoms, , https://www.youtube.com/watch?v=7WhRJV_bAiE

[Chem-periodic] Chemistry LibreTexts, 5.17: Electron Configurations and the Periodic Table, https://chem.libretexts.org/Bookshelves/General_Chemistry/Book%3A_ChemPRIME_(Moore_et_al.)/05%3A_The_Electronic_Structure_of_Atoms/5.17%3A_Electron_Configurations_and_the_Periodic_Table

[Chem_spectr] Chemistry LibreTexts, 7.3: The Atomic Spectrum of Hydrogen, https://chem.libretexts.org/Courses/Solano_Community_College/Chem_160/Chapter_07%3A_Atomic_Structure_and_Periodicity/7.03_The_Atomic_Spectrum_of_Hydrogen

[Clay] Millenium problems, https://www.claymath.org/millennium-problems

[Comp_Zoo] Complexity Zoo, https://complexityzoo.net/Complexity_Zoo

[Daskalatis] Costis Daskalakis, Equilibrium Computation & the Foundations of Deep Learning, Costis Daskalakis on Foundation of Data Science Series, Feb 18, 2021, https://www.youtube.com/watch?v=pDangP47ftE

[Deutsch-Jozsa] David Deutsch and Richard Jozsa, Rapid solutions of problems by quantum computation, Proceedings of the Royal Society of London A. 439: 553 558, `https://doi.org/10.1098/rspa.1992.0167`

[DiVincenzo] David P. DiVincenzo, The Physical Implementation of Quantum Computation, 10.1002/1521-3978(200009)48:9/11<771::AID-PROP771>3.0.CO;2-E, `https://arxiv.org/abs/quant-ph/0002077`

[Dirac_2] Dirac, P.A.M., The physical interpretation of the quantum dynamics, Proc. R. Soc. Lond. A 1927, 113, 621–641, `https://doi.org/10.1098/rspa.1927.0012`

[Dowling] Jonathan P. Dowling and Gerard J. Milburn, Quantum technology: the second quantum revolution, Royal Society, 20 June 2003, `https://doi.org/10.1098/rsta.2003.1227`

[E_mass] fundamental physical constants, electron mass, NIST, `https://physics.nist.gov/cgi-bin/cuu/Value?me|search_for=electron+mass`

[Fearnley] John Fearnley (University of Liverpool), Paul W. Goldberg (University of Oxford), Alexandros Hollender (University of Oxford), and Rahul Savani (University of Liverpool), The Complexity of Gradient Descent: CLS = PPAD ∩ PLS, STOC 2021: Proceedings of the 53rd Annual ACM SIGACT Symposium on Theory of Computing, June 2021 Pages 46–59, `https://doi.org/10.1145/3406325.3451052`

[Getty] Girl looking up, `https://media.gettyimages.com/photos/you-learn-something-new-every-day-picture-id523149221?k=20&m=523149221&s=612x612&w=0&h=7ZFg6ETuKlqr1nzi98IBNz-uYXccQwiuNKEk0hGKKIU=`

[Hill] Learning Scientific Programming with Python, Chapter 2: The Core Python Language I, Problems, P2.5, Electronic configurations, `https://scipython.com/book/chapter-2-the-core-python-language-i/questions/problems/p25/electronic-configurations/`

[Intro_BOA_1] M. Born, J.R. Oppenheimer, On the Quantum theory of molecules, `https://www.theochem.ru.nl/files/dbase/born-oppenheimer-translated-s-m-blinder.pdf`

[Intro_BOA_2] M. Born and R. J. Oppenheimer, Zur Quantentheorie der Molekeln, Annalen der physik, 20, 457-484 (August 1927), `https://doi.org/10.1002/andp.19273892002`

[Kaplan] Ilya G. Kaplan, Modern State of the Pauli Exclusion Principle and the Problems of Its Theoretical Foundation, Symmetry 2021, 13(1), 21, `https://doi.org/10.3390/sym13010021`

[Knill] Emanuel Knill, Raymond Laflamme, A Theory of Quantum Error-Correcting Codes, https://arxiv.org/abs/quant-ph/9604034

[Lyman_series] Lyman series, From Wikipedia, https://en.wikipedia.org/wiki/Lyman_series

[MIT_QC_1981] MIT Endicott House, The Physics of Computation Conference, Image "Physics of Computation Conference, Endicott House MIT May 6-8, 1981", Mar 21, 2018, https://mitendicotthouse.org/physics-computation-conference/

[Neutron-electron-mass-ratio] neutron-electron mass ratio, NIST, https://physics.nist.gov/cgi-bin/cuu/Value?mnsme

[Orb_Approx] Definition of Orbital Approximation, https://www.chemicool.com/definition/orbital-approximation.html

[Part_1] List of particles, Wikipedia, https://en.wikipedia.org/wiki/List_of_particles

[Pauling] L. Pauling and E. B. Wilson, Introduction to Quantum Mechanics with Applications to Chemistry, Dover (1935)

[PoorLeno] File:Hydrogen Density Plots.png, From Wikipedia, https://en.wikipedia.org/wiki/File:Hydrogen_Density_Plots.png

[Preskill_40y] John Preskill, Quantum computing 40 years later, https://arxiv.org/abs/2106.10522

[PvsNP] P and NP, www.cs.uky.edu. Archived from the original on 2016-09-19, https://web.archive.org/web/20160919023326/http://www.cs.uky.edu/~lewis/cs-heuristic/text/class/p-np.html

[QC40] (Livestream) QC40: Physics of Computation Conference 40th Anniversary, https://www.youtube.com/watch?v=GR6ANm6Z0yk

[QMC] Google Quantum AI, Unbiased fermionic Quantum Monte Carlo with a Quantum computer, Quantum Summer Symposium 2021, 30 July 2021, https://www.youtube.com/watch?v=pTHtyKuByvw

[Qa_Zoo] Stephen Jordan, Algebraic and Number Theoretic Algorithms, https://quantumalgorithmzoo.org/

[Rayleigh_Ritz] Rayleigh-Ritz method, Wikipedia, https://en.wikipedia.org/wiki/Rayleigh%E2%80%93Ritz_method

[Rydberg_R] Rydberg constant, Wikipedia, `https://en.wikipedia.org/wiki/Rydberg_constant`

[Rydberg_Ritz] Rydberg-Ritz combination principle, Wikipedia, `https://en.wikipedia.org/wiki/Rydberg%E2%80%93Ritz_combination_principle`

[Shor] Peter Shor, The Story of Shor's Algorithm, Straight From the Source, July 2, 2021, `https://www.youtube.com/watch?v=6qD9XElTpCE`

[VQE_1] Peruzzo, A., McClean, J., Shadbolt, P. et al., A variational eigenvalue solver on a photonic quantum processor, Nat Commun 5, 4213 (2014), `https://doi.org/10.1038/ncomms5213`

[VQE_2] Qiskit Nature, Ground state solvers, `https://qiskit.org/documentation/nature/tutorials/03_ground_state_solvers.html`

[Vandersypen] Vandersypen, L., Steffen, M., Breyta, G. et al., Experimental realization of Shor's quantum factoring algorithm using nuclear magnetic resonance, Nature 414, 883–887 (2001), `https://doi.org/10.1038/414883a`

2

Postulates of Quantum Mechanics

"The vivid force of his mind prevailed, and he fared forth far beyond the flaming ramparts of the heavens and traversed the boundless universe in thought and mind."

– Titus Lucretius Carus

Figure 2.1 – Titus Lucretius Carus gazing at the Milky Way galaxy [authors] built from an image of Titus Lucretius in the public domain and an image of the Milky Way galaxy [NASA]

In the first two books of his six-book poem De Rerum Natura (On the Nature of Things), Titus Lucretius Carus, a Roman poet and philosopher, discusses life and love and explains the basic principles of Epicurean physics, a Greek way of understanding the world before Christ [Lucr_1]. He put forward the idea that matter is both active and indeterminate [Lucr_2], a very "quantum" way of thinking to say the least.

Using an analogy of dust particles in a sunbeam, Lucretius described what is now known as Brownian motion [Lucr_3]. He talked about matter and used concepts such as mostly empty space to describe it. It would take more than 2 millennia for these ideas to become widely adopted and put into the postulates of quantum mechanics. We reviewed the milestones of the late 1800s and early 1900s that lead to the postulates of quantum mechanics in *Chapter 1, Introducing Quantum Concepts*.

The five postulates of quantum mechanics are not considered the law of nature and cannot be shown to be true, neither mathematically nor experimentally. Rather, the postulates are simply guidelines for the behavior of particles and matter. Even though it took a few decades for the postulates to be formulated and a century to be utilized by the broader scientific community, the postulates remain a powerful tool for predicting the properties of matter and particles and are the foundation of quantum chemistry and computing.

In this chapter, we will cover the following topics:

- *Section 2.1, Postulate 1 – Wave functions*
- *Section 2.2, Postulate 2 – Probability amplitudes*
- *Section 2.3, Postulate 3 – Measurable quantities and operators*
- *Section 2.4, Postulate 4 – Time independent stationary states*
- *Section 2.5, Postulate 5 – Time evolution dynamics, Schrödinger's equation*

In this chapter, we primarily focus on the significance of Postulate 1, Wave functions, because we think that this postulate has powerful repercussions for useful innovations. Traditionally, Postulate 1 is hard to grasp conceptually and has been a scientific challenge to represent mathematically and artistically. We have taken active steps to overcome this artistically, as shown in *Figure 1.4* and in *Figure 2.2*. The other four postulates support Postulate 1. We do not go into as much detail with these postulates as we do with Postulate 1 in this chapter; however, will be utilizing them in subsequent chapters. Readers who are not familiar with linear algebra or with Dirac notation are invited to refer to *Appendix A – Readying Mathematical Concepts*.

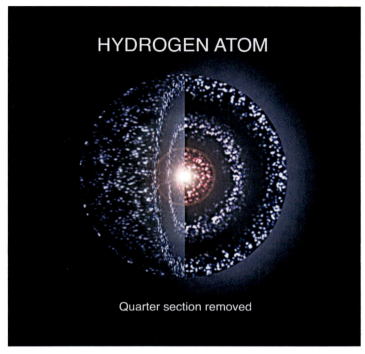

Figure 2.2 – Artistic image of a hydrogen atom wave function [authors]

Technical requirements

A companion Jupyter notebook for this chapter can be downloaded from GitHub at `https://github.com/PacktPublishing/Quantum-Chemistry-and-Computing-for-the-Curious`, which has been tested in the Google Colab environment, which is free and runs entirely in the cloud, and in the IBM Quantum Lab environment. Please refer to *Appendix B – Leveraging Jupyter Notebooks in the Cloud*, for more information. The companion Jupyter notebook automatically installs the following list of libraries:

- **Numerical Python** (**NumPy**) [NumPy], an open-source Python library that is used in almost every field of science and engineering
- **SymPy**, [SymPy] a Python library for symbolic mathematics
- **Qiskit** [Qiskit], an open-source SDK for working with quantum computers at the level of pulses, circuits, and application modules
- Qiskit visualization support to enable the use of its visualization functionality and Jupyter notebooks

Install NumPy using the following command:

```
pip install numpy
```

Install SymPy using the following command:

```
pip install sympy
```

Install Qiskit using the following command:

```
pip install qiskit
```

Install Qiskit visualization support using the following command:

```
pip install 'qiskit[visualization]'
```

Import math libraries using the following commands:

```
import cmath
import math
```

2.1. Postulate 1 – Wave functions

The total wave function describes the physical behavior of a system and is represented by the capital Greek letter Psi: Ψ_{total}. It contains all the information of a quantum system and includes complex numbers ($z = a + ib$) as parameters. In general, Ψ_{total} is a function of all the particles in the system $\{1, ..., i, ..., N\}$, where the total number of particles is N. Furthermore, Ψ_{total} includes the spatial position of each particle ($r_i = \{x_i, y_i, z_i\}$), the spin directional coordinates for each particle ($s_i = \{s_{x_i}, s_{y_i}, s_{z_i}\}$), and time ($t$):

$$\Psi_{total}(r, s, t)$$

where r and s are vectors of single-particle coordinates:

$$r = \{r_1, ..., r_i, ..., r_N\}$$
$$s = \{s_1, ..., s_i, ..., s_N\}$$

The total wave function for a one-particle system is a product of a spatial $\psi(r_1)$, spin $\chi(s_1)$, and time $f(t)$ functions:

$$\Psi_{total}(r_1, s_1, t) = \psi(r_1) * \chi(s_1) * f(t)$$

If the wave function for a multiple-particle system cannot be factored into a product of single-particle functions, then we consider the quantum system as **entangled**. If the wave function can be factored into a product of single-particle functions, then it is not entangled and is called a separable state. We will revisit the concept of entanglement in *Chapter 3, Quantum Circuit Model of Computation.*

The spatial part of the wave function $\psi(r)$ can be converted from Cartesian coordinates (x, y, z) to spherical coordinates (r, θ, φ) where r is the radial distance determined by the distance formula $\sqrt{x^2 + y^2 + z^2}$, θ is the polar angle ranging from 0 to π ($\theta \in [0,\pi]$), and φ is the azimuthal angle ranging from 0 to 2π ($\varphi \in [0, 2\pi]$), through the following equations:

- $x = r \sin \theta \, \cos \varphi$

- $y = r \sin \theta \sin \varphi$

- $z = r \cos \theta$

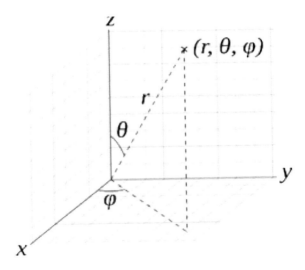

Figure 2.3 – Spherical coordinates [public domain]

There are certain properties of a wave function that need to be properly considered in order to accurately represent a quantum system:

- **Single-valued,** meaning that for a given input variable there is only one possible output

- **Positive definite,** meaning that the complex conjugate transpose of the wave function, indicated by a dagger (†), times the wave function itself is strictly greater than zero: $\Psi_{total}^{\dagger} \Psi_{total} > 0$

- **Square integrable**, meaning that the positive definite product is less than infinity when integrated over all space $(\boldsymbol{r})$: $\int_{-\infty}^{\infty} \Psi_{total}^{\dagger}(\boldsymbol{r})\Psi_{total}(\boldsymbol{r})\, d\boldsymbol{r} < \infty$, where $d\boldsymbol{r} = dx\, dy\, dz$

- **Normalizable**, meaning that a particle must exist in a volume (τ) and at a point in time, that is, it must exist somewhere in all space and time:

$\int_{-\infty}^{\infty} \Psi_{total}^{\dagger}(\boldsymbol{r}, t)\Psi_{total}(\boldsymbol{r}, t)\, d\tau\, dt = 1$, where $d\tau = r^2 \sin\theta\, dr\, d\theta\, d\varphi$

- **Complete**, meaning that all statistically important data that is needed to represent that quantum system is available such that calculations of properties converge to a limit, that is, a single value

For quantum chemistry applications, we will use Python code to show how to include the spatial $\psi(\boldsymbol{r})$ and spin $\chi(\boldsymbol{s})$ functions:

- *Section 2.1.1, Spherical harmonic functions*, which are related to the quantum numbers n, l, m_l and to the spatial variables x, y, z or r, θ, φ

- *Section 2.1.2, Addition of momenta using Clebsch-Gordan (CG) coefficients*, which is for coupling multiple particles and can be applied to both orbital (l, m_l) and spin quantum numbers (s, m_s)

- *Section 2.1.3, The general formulation of the Pauli exclusion principle*, which ensures the proper symmetry requirements for a multiple particle system: either totally fermionic, totally bosonic, or a combination of the two

From a machine learning perspective, there are other parameters that the wave function can depend on. These parameters are called hyperparameters and are used to optimize the wave function to obtain the most accurate picture of the state of interest.

2.1.1. Spherical harmonic functions

Spherical harmonic functions $\left(Y_l^{m_l}\right)$ are used to describe one-electron systems and depend on the angular momentum (l) and the magnetic quantum number (m_l), as well as the spatial coordinates:

$$\psi_{lm_l}(x, y, z) \approx \psi_{lm_l}(r, \theta, \varphi)$$

and are a set of special functions defined on the surface of a sphere called the radial wave function $R(r)$:

$$\psi_{nlm_l}(r, \theta, \varphi) = R(r)\, Y_l^{m_l}(\theta, \varphi)$$

Since the hydrogen atom is the simplest atom, consisting of only one electron around a single proton, in this section we will illustrate what these functions look like. Some of the spherical harmonic functions for the hydrogen atom are shown in *Figure 2.4*:

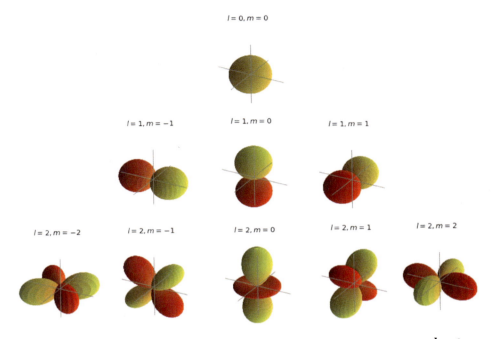

Figure 2.4 – Spatial wave functions of the hydrogen atom with quantum numbers l and m_l

Recall the following:

- The principal quantum (n) is a continuous quantum variable that ranges from 1 to infinity such that in practice, due to ionization, it becomes a discrete variable.

- The angular momentum quantum number (l) is contained in the discrete set determined: $l \in \{0, \ldots, n-1\}$.

- The magnetic quantum number (m_l) is contained in the discrete set determined by the angular momentum quantum number (l): $m_l \in \{-l, \ldots, 0, \ldots, l\}$.

The spherical harmonic functions, $Y_l^{m_l}(\theta, \varphi)$, can be split into a product of three functions:

$$Y_l^{m_l}(\theta, \varphi) = A(l, m_l)\, P_l^{m_l}(\cos \theta)\, S_{m_l}(\varphi),$$

where $A(l, m_l)$ is a constant that depends only on the quantum numbers (l, m_l), $P_l^{m_l}(\cos \theta)$ is a polar function, also known as the associated Legendre polynomial functions, which can be a complex function if the angular momentum (l) is positive or negative, and $S_{m_l}(\varphi) = e^{im_l\varphi}$ is a complex exponential azimuthal function. To illustrate spherical harmonic functions, we use the following code, which computes them [SciPy_sph], then casts them into the following real functions [Sph_Real]:

$$Y_l^{m_l} = \begin{cases} \sqrt{2}\,(-1)^{m_l}\, Im\left[Y_l^{|m_l|}\right], & m_l < 0 \\ Y_l^0, & m_l = 0 \\ \sqrt{2}\,(-1)^{m_l}\, Re\left[Y_l^{m_l}\right], & m_l > 0 \end{cases}$$

and finally displays these real functions in three dimensions with Python's Matplotlib module. Let's now implement this in Python.

Importing NumPy, SciPy, and Matplotlib Python modules

The following Python statements import the required NumPy, SciPy, and Matplotlib modules:

```
import numpy as np
import matplotlib.pyplot as plt
import matplotlib.gridspec as gridspec
from scipy.special import sph_harm
```

Setting-up grids of polar (theta − θ) and azimuthal (phi − φ) angles

We define a function called `setup_grid()` that creates a grid of polar coordinates and the corresponding cartesian coordinates with the following Python functions:

- `numpy.linspace`: Returns evenly spaced numbers over a specified interval

- `numpy.meshgrid`: Returns coordinate matrices from coordinate vectors

The `setup_grid()` function has one input parameter, `num`, which is a positive integer that is the number of distinct values of polar coordinates.

It returns the following:

- `theta`, `phi`: Two-dimensional NumPy arrays of shape `num` x `num`
- `xyz`: Three-dimensional NumPy array of shape `(3, num, num)`

```
def setup_grid(num=100):
    theta = np.linspace(0, np.pi, num)
    phi = np.linspace(0, 2*np.pi, num)
    # Create a 2D meshgrid from two 1D arrays of theta, phi
coordinates
    theta, phi = np.meshgrid(theta, phi)
    # Compute cartesian coordinates with radius r = 1
    xyz = np.array([np.sin(theta) * np.sin(phi),
                    np.sin(theta) * np.cos(phi),
                    np.cos(theta)])
    return (theta, phi, xyz)
```

Let's check the shape of the NumPy arrays returned by `setup_grid()`:

```
(theta, phi, xyz) = setup_grid()
print("Shape of meshgrid arrays, theta: {}, phi: {}, xyz: {}".
format(theta.shape, phi.shape, xyz.shape))
```

Here's the output:

```
Shape of meshgrid arrays, theta: (100, 100), phi: (100, 100),
xyz: (3, 100, 100)
```

Coloring the plotted surface of the real functions of the spherical harmonic function (Y)

We define a function called `colour_plot()` that colors the plotted surface of the real functions of the spherical harmonic $Y_l^{m_l}$ according to the sign of its real part, $Re\left[Y_l^{m_l}\right]$. It has the following input parameters:

- `ax`: A three-dimensional Matplotlib figure
- `Y`: A spherical harmonic function
- `Yx,Yy,Yz`: Cartesian coordinates of the plotted surface of the spherical harmonic function

- cmap: A built-in colormap accessible via the `matplotlib.cm.get_cmap` function [Cmap], for instance, autumn, cool, spring, and winter:

```
def colour_plot(ax, Y, Yx, Yy, Yz, cmap):
    # Colour the plotted surface according to the sign of Y.real
    # https://matplotlib.org/stable/gallery/mplot3d/surface3d.html?highlight=surface%20plots
    # https://matplotlib.org/stable/tutorials/colors/colormaps.html
    cmap = plt.cm.ScalarMappable(cmap=plt.get_cmap(cmap))
    cmap.set_clim(-0.5, 0.5)
    ax.plot_surface(Yx, Yy, Yz,
                    facecolors=cmap.to_rgba(Y.real),
                    rstride=2, cstride=2)
    return
```

Defining a function that plots a set of x, y, z axes and sets the title of a figure

We define a function called `draw_axes()` that plots the axes of a Matplotlib figure and sets a title. It has three input parameters:

1. `ax`: A three-dimensional Matplotlib figure
2. `ax_lim`: A positive real number that controls the size of the plotted surface
3. `title`: A string of characters that will be shown as the title of the output figure:

```
def draw_axes(ax, ax_lim, title):
    ax.plot([-ax_lim, ax_lim], [0,0], [0,0], c='0.5', lw=1, zorder=10)
    ax.plot([0,0], [-ax_lim, ax_lim], [0,0], c='0.5', lw=1, zorder=10)
    ax.plot([0,0], [0,0], [-ax_lim, ax_lim], c='0.5', lw=1, zorder=10)

    # Set the limits, set the title and then turn off the axes frame
    ax.set_title(title)
    ax.set_xlim(-ax_lim, ax_lim)
    ax.set_ylim(-ax_lim, ax_lim)
```

```
ax.set_zlim(-ax_lim, ax_lim)
ax.axis('off')
return
```

Defining a function that computes the real form of the spherical harmonic function (Y)

Please be cautious in this part of the code because SciPy defines theta (θ) as the azimuthal angle and phi (φ) as the polar angle [SciPy_sph], which is opposite of the standard definitions used for plotting.

The comb_Y() function takes the following input parameters:

- l: Angular momentum quantum number
- m: Magnetic quantum number m_l
- theta, phi: Two-dimensional NumPy arrays of shape num x num

It returns the real form of the spherical harmonic function $Y_l^{m_l}$ presented earlier:

```
def comb_Y(l, m, theta, phi):
    Y = sph_harm(abs(m), l, phi, theta)
    if m < 0:
        Y = np.sqrt(2) * (-1)**m * Y.imag
    elif m > 0:
        Y = np.sqrt(2) * (-1)**m * Y.real
    return Y
```

Defining a function that displays the spatial wave functions for a range of values of the angular momentum quantum number and the magnetic quantum number

The following function displays spatial wave functions for l in range $[0, k]$, where k is a parameter and m_l is in range $[-l, l]$ as illustrated in the following code for the hydrogen atom in states $l = 0, l = 1, l = 2$:

```
def plot_orbitals(k, cmap = 'autumn'):
    for l in range(0, k+1):
        for m in range(-l, l+1):
            fig = plt.figure(figsize=plt.figaspect(1.))
            (theta, phi, xyz) = setup_grid()
```

```
        ax = fig.add_subplot(projection='3d')
        Y = comb_Y(l, m, theta, phi)
        title = r'$l={{{}}}, m={{{}}}$'.format(l, m)
        Yx, Yy, Yz = np.abs(Y) * xyz
        colour_plot(ax, Y, Yx, Yy, Yz, cmap)
        draw_axes(ax, 0.5, title)
        fig_name = 'Hydrogen_l'+str(l)+'_m'+str(m)
        plt.savefig(fig_name)
        plt.show()
    return
```

Spatial wave functions of the hydrogen atom

The spatial wave functions for the one electron of the hydrogen atom in states $l = 0, l = 1, l = 2$, and m_l in range $[-l, l]$ are computed and displayed with the `plot_orbitals` Python function defined earlier:

```
 plot_orbitals(2)
```

The result is shown in *Figure 2.4*.

Questions to consider

What happens to the spherical harmonic functions when we have more than one electron, that is, in heavier elements? How do these functions operate or change? For instance, what happens when there are three electrons with non-zero angular momentum, as in the case with the nitrogen atom?

To accomplish this kind of complexity and variability, we need to add or couple angular momentum using **Clebsch-Gordon (CG) coefficients**, as presented in *Section 2.1.2, Addition of momenta using CG coefficients*.

2.1.2. Addition of momenta using CG coefficients

The addition or coupling of two momenta (j_1 and j_2) along with the associated projections (m_{j_1} and m_{j_2}) is described by the summation of two initial state wave functions $|j_1, m_{j_1}\rangle$ and $|j_2, m_{j_2}\rangle$ over the possible or allowed quantum numbers:

$$|j_{1,2}, m_{j_{1,2}}\rangle = \sum_{m_{j_1}=-j_1,\ m_{j_2}=-j_2}^{j_1,\ j_2} \langle j_1, m_{j_1}, j_2, m_{j_2} | j_{1,2}, m_{j_{1,2}} \rangle |j_1, m_{j_1}\rangle |j_2, m_{j_2}\rangle$$

to a final state wave function of choice $\left|j_{1,2}, m_{j_{1,2}}\right\rangle$. Yes, we can choose the final state as we please, if we follow the rules of vector addition. The CG coefficients are the expansion coefficients of coupled total angular momentum in an uncoupled tensor product basis:

$$C_{j_1,m_{j_1},j_2,m_{j_2}}^{j_{1,2},m_{j_{1,2}}} = \left\langle j_1, m_{j_1}, j_2, m_{j_2} \middle| j_{1,2}, m_{j_{1,2}} \right\rangle$$

We use a generic j and m_j to represent a formula where either angular (l and m_l) and/or spin (s and m_s) momentum can be coupled together. We can couple angular momenta only, or spin momenta only, or the two together. The addition is accomplished by knowing the allowed values for the quantum numbers.

Using CG coefficients with Python SymPy

The Python SymPy library [SymPy_CG] implements the formula with the CG class as follows.

Class $CG(j_1, m_{j_1}, j_2, m_{j_2}, j_{1,2}, m_{j_{1,2}})$ has the following parameters:

- j_1, m_{j_1}: Angular momentum and the projection of state 1
- j_2, m_{j_2}: Angular momentum the projection of state 2
- $j_{1,2}, m_{1,2}$: Total angular momentum of the coupled system

Importing the SymPy CG coefficients module

The following statements import the SymPy CG coefficients module:

```
import sympy
from sympy import S
from sympy.physics.quantum.cg import CG, cg_simp
```

Defining a CG coefficient and evaluating its value

We can couple two electrons (fermions) in a spin paired state in two different ways: symmetric or antisymmetric. We denote spin up in the z direction as $\left|s_1 = \frac{1}{2}, m_{s_1} = \frac{1}{2}\right\rangle = \uparrow$ and spin down in the z direction as $\left|s_1 = \frac{1}{2}, m_{s_1} = -\frac{1}{2}\right\rangle = \downarrow$.

We can also couple spin states with angular momentum states. When coupling angular momentum (l) with spin (s), we change the notation to j. We go through these three examples next.

Fermionic spin pairing to symmetric state ($s_{1,2} = 1, m_{s_{1,2}} = 0$)

The coupling of the symmetric spin paired state is $|s_{1,2} = 1, m_{s_{1,2}} = 0\rangle$ and is described by the following equation:

$$|s_{1,2} = 1, m_{s_{1,2}} = 0\rangle =$$

$$\langle s_1 = \tfrac{1}{2}, m_{s_1} = \tfrac{1}{2}, s_2 = \tfrac{1}{2}, m_{s_2} = -\tfrac{1}{2}|s_{1,2} = 1, m_{s_{1,2}} = 0\rangle |s_1 = \tfrac{1}{2}, m_{s_1} = \tfrac{1}{2}\rangle |s_2 = \tfrac{1}{2}, m_{s_2} = -\tfrac{1}{2}\rangle$$

$$+ \langle s_1 = \tfrac{1}{2}, m_{s_1} = -\tfrac{1}{2}, s_1 = \tfrac{1}{2}, m_{s_1} = \tfrac{1}{2}|s_{1,2} = 1, m_{s_{1,2}} = 0\rangle |s_1 = \tfrac{1}{2}, m_{s_1} = -\tfrac{1}{2}\rangle |s_1 = \tfrac{1}{2}, m_{s_1} = \tfrac{1}{2}\rangle$$

Using the following code, we obtain the CG coefficients for the preceding equation:

```
CG(S(1)/2, S(1)/2, S(1)/2, -S(1)/2, 1, 0).doit()
CG(S(1)/2, -S(1)/2, S(1)/2, S(1)/2, 1, 0).doit()
```

Here's the result:

$$\frac{\sqrt{2}}{2}, \frac{\sqrt{2}}{2}$$

Figure 2.5 – Defining a CG coefficient and evaluating its value

Plugging in the CG coefficients as well as the up-spin and down-spin functions, we get this:

$$|s_{1,2} = 1, m_{s_{1,2}} = 0\rangle$$
$$= \frac{1}{\sqrt{2}}\left\{|s_1 = \tfrac{1}{2}, m_{s_1} = \tfrac{1}{2}\rangle |s_2 = \tfrac{1}{2}, m_{s_2} = -\tfrac{1}{2}\rangle + |s_1 = \tfrac{1}{2}, m_{s_1} = -\tfrac{1}{2}\rangle |s_2 = \tfrac{1}{2}, m_{s_2} = \tfrac{1}{2}\rangle\right\}$$

$$|s_{1,2} = 1, m_{s_{1,2}} = 0\rangle = \frac{1}{\sqrt{2}}\{\uparrow\downarrow + \downarrow\uparrow\}$$

Fermionic spin pairing to antisymmetric state ($s_{1,2} = 0, m_{s_{1,2}} = 0$)

The coupling of the antisymmetric spin paired state, $|s_{1,2} = 0, m_{s_{1,2}} = 0\rangle$, is described by the following equation:

$$|s_{1,2} = 0, m_{s_{1,2}} = 0\rangle =$$

$$\langle s_1 = \tfrac{1}{2}, m_{s_1} = \tfrac{1}{2}, s_2 = \tfrac{1}{2}, m_{s_2} = -\tfrac{1}{2}|s_{1,2} = 0, m_{s_{1,2}} = 0\rangle |s_1 = \tfrac{1}{2}, m_{s_1} = \tfrac{1}{2}\rangle |s_2 = \tfrac{1}{2}, m_{s_2} = -\tfrac{1}{2}\rangle$$

$$+ \langle s_1 = \tfrac{1}{2}, m_{s_1} = -\tfrac{1}{2}, s_1 = \tfrac{1}{2}, m_{s_1} = \tfrac{1}{2}|s_{1,2} = 0, m_{s_{1,2}} = 0\rangle |s_1 = \tfrac{1}{2}, m_{s_1} = -\tfrac{1}{2}\rangle |s_1 = \tfrac{1}{2}, m_{s_1} = \tfrac{1}{2}\rangle$$

Using the following code, we obtain the CG coefficients for the preceding equation:

```
CG(S(1)/2, S(1)/2, S(1)/2, -S(1)/2, 0, 0).doit()
CG(S(1)/2, -S(1)/2, S(1)/2, S(1)/2, 0, 0).doit()
```

Here's the result:

$$\frac{\sqrt{2}}{2}, -\frac{1}{\sqrt{2}}$$

Figure 2.6 – Defining a CG coefficient and evaluating its value

Plugging in the CG coefficients as well as the up-spin and down-spin functions, we get the following:

$$|s_{1,2} = 0, m_{s_{1,2}} = 0\rangle$$

$$= \frac{1}{\sqrt{2}}\left\{\left|s_1 = \frac{1}{2}, m_{s_1} = \frac{1}{2}\right\rangle\left|s_2 = \frac{1}{2}, m_{s_2} = -\frac{1}{2}\right\rangle - \left|s_1 = \frac{1}{2}, m_{s_1} = -\frac{1}{2}\right\rangle\left|s_2 = \frac{1}{2}, m_{s_2} = \frac{1}{2}\right\rangle\right\}$$

$$|s_{1,2} = 0, m_{s_{1,2}} = 0\rangle = \frac{1}{\sqrt{2}}\{\uparrow\downarrow - \downarrow\uparrow\}$$

Coupling spin and angular momentum ($j_{1,2} = \frac{1}{2}, j_{s_{1,2}} = \frac{1}{2}$)

Let's couple together angular momenta with $l_1 = 1$ and $m_{l_1} = \{1, 0, -1\}$ to a fermionic spin state $s_2 = \frac{1}{2}$ and $m_{s_2} = \{\frac{1}{2}, -\frac{1}{2}\}$ for a final state of choice of $\left|j_{1,2} = \frac{1}{2}, j_{s_{1,2}} = \frac{1}{2}\right\rangle$:

$$\left|j_{1,2} = \frac{1}{2}, j_{s_{1,2}} = \frac{1}{2}\right\rangle =$$

$$\left\langle l_1 = 1, m_{l_1} = 0, s_2 = \frac{1}{2}, m_{s_2} = \frac{1}{2}\middle|j_{1,2} = \frac{1}{2}, m_{j_{1,2}} = \frac{1}{2}\right\rangle|l_1 = 1, m_{l_1} = 0\rangle\left|s_2 = \frac{1}{2}, m_{s_2} = \frac{1}{2}\right\rangle$$

$$+ \left\langle l_1 = 1, m_{l_1} = 1, s_2 = \frac{1}{2}, m_{s_2} = -\frac{1}{2}\middle|j_{1,2} = \frac{1}{2}, m_{j_{1,2}} = \frac{1}{2}\right\rangle|l_1 = 1, m_{l_1} = 1\rangle\left|s_2 = \frac{1}{2}, m_{s_2} = -\frac{1}{2}\right\rangle$$

$$+ \left\langle l_1 = 1, m_{l_1} = -1, s_2 = \frac{1}{2}, m_{s_2} = \frac{1}{2}\middle|j_{1,2} = \frac{1}{2}, m_{j_{1,2}} = \frac{1}{2}\right\rangle|l_1 = 1, m_{l_1} = -1\rangle\left|s_2 = \frac{1}{2}, m_{s_2} = \frac{1}{2}\right\rangle.$$

The CG coefficients of this equation are calculated using the following code:

```
CG(1, 0, S(1)/2, S(1)/2, S(1)/2, S(1)/2).doit()
CG(1, 1, S(1)/2, -S(1)/2, S(1)/2, S(1)/2).doit()
CG(1, -1, S(1)/2, S(1)/2, S(1)/2, S(1)/2).doit()
```

Here's the result:

$$-\frac{\sqrt{3}}{3},\frac{\sqrt{6}}{3},0$$

Figure 2.7 – Defining a CG coefficient and evaluating its value

Plugging the result of the preceding code into the formula, we obtain the following:

$$\left|j_{1,2}=\frac{1}{2},j_{s_{1,2}}=\frac{1}{2}\right\rangle=$$

$$-\frac{\sqrt{3}}{3}\left|l_1=1,m_{l_1}=0\right\rangle\left|s_2=\frac{1}{2},m_{s_2}=\frac{1}{2}\right\rangle$$

$$+\frac{\sqrt{6}}{3}\left|l_1=1,m_{l_1}=1\right\rangle\left|s_2=\frac{1}{2},m_{s_2}=-\frac{1}{2}\right\rangle$$

$$+0\left|l_1=1,m_{l_1}=-1\right\rangle\left|s_2=\frac{1}{2},m_{s_2}=\frac{1}{2}\right\rangle.$$

Now, we reduce this and plug in the up-spin and down-spin functions:

$$\left|j_{1,2}=\frac{1}{2},j_{s_{1,2}}=\frac{1}{2}\right\rangle=\frac{\sqrt{3}}{3}\{-\left|l_1=1,m_{l_1}=0\right\rangle\uparrow+\sqrt{2}\left|l_1=1,m_{l_1}=1\right\rangle\downarrow\}.$$

In the last step, we plugged in the spherical harmonic functions for the following:

- $\left|l_1=1,m_{l_1}=0\right\rangle=\frac{1}{2}\frac{\sqrt{3}}{\sqrt{\pi}}\cos\theta$

- $\left|l_1=1,m_{l_1}=1\right\rangle=-\frac{1}{2}\frac{\sqrt{3}}{\sqrt{2\pi}}e^{-i\varphi}\sin\theta$

In doing so, we obtain this:

$$\left|j_{1,2}=\frac{1}{2},j_{s_{1,2}}=\frac{1}{2}\right\rangle=-\frac{\sqrt{3}}{3}\left\{\frac{1}{2}\frac{\sqrt{3}}{\sqrt{\pi}}\cos\theta\uparrow+\frac{\sqrt{2}}{2}\frac{\sqrt{3}}{\sqrt{2\pi}}e^{-i\varphi}\sin\theta\downarrow\right\}=-\frac{1}{2}\frac{1}{\sqrt{\pi}}\{\cos\theta\uparrow+e^{-i\varphi}\sin\theta\downarrow\}.$$

Then we can drop the factor of $-\frac{1}{2}\frac{1}{\sqrt{\pi}}$, as it is a global factor, so that the final state is as follows:

$$\left|j_{1,2}=\frac{1}{2},j_{s_{1,2}}=\frac{1}{2}\right\rangle=\{\cos\theta\uparrow+e^{-i\varphi}\sin\theta\downarrow\}.$$

Some of you might recognize this function as a qubit wave function for computing without including time dependence. In fact, for the state of a qubit, we change the up arrow (↑) to ket 0 ($|0\rangle$) to indicate the magnetic projection of zero ($m_{l_1} = 0$), and likewise for the down arrow (↓) to ket 1 ($|1\rangle$) to indicate the magnetic projection of zero ($m_1 = 1$). With this, we have the following:

$$\left| j_{1,2} = \frac{1}{2}, j_{s_{1,2}} = \frac{1}{2} \right\rangle = \{\cos\theta\, |0\rangle + e^{-i\varphi} \sin\theta\, |1\rangle\}.$$

We cover this topic in more detail in *Chapter 3, Quantum Circuit Model of Computation*.

Spatial wave functions of different states of the nitrogen atom with three p electrons

Now we would like to illustrate the wave function of the nitrogen atom with three p electrons [Sharkey_0]. We chose this system because we are coupling more than two non-zero momentum vectors by expressing its coupled total momentum in an uncoupled tensor product basis of each electron [Phys5250]. This means that we assumed the wave function is not entangled. We must apply the addition of angular momenta formula twice (recursively) so that we have all the combinations of coupling with the final state of choice. The different shapes of the spatial wave function of the nitrogen atom with three p electrons are shown here:

Figure 2.8 – Spatial wave functions of different states of the nitrogen atom with three p electrons

We will go through the example for the final state of $L = 0, M = 0$.

Spatial wave function of the ground state of the nitrogen atom with 3 p electrons in $L = 0, M = 0$

Electrons are fermions, and therefore they cannot occupy the same set of quantum numbers. Because we are working with three p electrons, the orbital angular momentum (l) for each electron is as follows:

$$l_1 = 1, l_2 = 1, \text{ and } l_3 = 1.$$

This couples with the final momentum state of $l_{123}(L) = 0$. The allowable set of magnetic momenta (m_l) for each electron is as follows:

$$m_{l_1} = \{1, 0, -1\}, m_{l_2} = \{1, 0, -1\}, \text{ and } m_{l_3} = \{1, 0, -1\},$$

and the final coupled magnetic projection state is:

$$m_{l_{123}}(M) = 0$$

To accomplish this type of coupling of three momenta, we must apply the addition of angular momenta formula twice (recursively) so that we have all the combinations of coupling with the final state, $|L = 0, M = 0\rangle$.

Each electron is in the same shell or principal quantum number (n) level; however, each is in a different subshell (m_l) and has a spin of either up or down. For this example, the spin state is irrelevant, and we are choosing to not include it. Since these electrons are in different subshells, that means they cannot have the same combination of quantum numbers (l and m_l):

	$|l_1, m_{l_1}\rangle$	$|l_2, m_{l_2}\rangle$	$|l_3, m_{l_3}\rangle$
0	$|1, -1\rangle$	$|1, 0\rangle$	$|1, 1\rangle$
1	$|1, -1\rangle$	$|1, 1\rangle$	$|1, 0\rangle$
2	$|1, 0\rangle$	$|1, -1\rangle$	$|1, 1\rangle$
3	$|1, 0\rangle$	$|1, 1\rangle$	$|1, -1\rangle$
4	$|1, 1\rangle$	$|1, -1\rangle$	$|1, 0\rangle$
5	$|1, 1\rangle$	$|1, 0\rangle$	$|1, -1\rangle$

Figure 2.9 – Electron configurations of $L = 0, M = 0$

Setting up a dictionary of six configuration tuples

Each tuple contains $l_1, m_{l_1}, l_2, m_{l_2}, l_{12}, m_{l_{12}}, l_3, m_{l_3}, l_{123}(L), m_{123}(M)$, where $l_{12}, m_{l_{12}}$ is the first coupling from electron 1 with 2, and $, l_{123}(L), m_{123}(M)$ is the second coupling of electrons 1 and 2 with 3:

```
T00 = {0:  (1,-1,  1,0,   1,-1,  1,1,    0,0),
       1:  (1,-1,  1,1,   1,0,   1,0,    0,0),
       2:  (1,0,   1,-1,  1,-1,  1,1,    0,0),
       3:  (1,0,   1,1,   1,1,   1,-1,   0,0),
       4:  (1,1,   1,-1,  1,0,   1,0,    0,0),
       5:  (1,1,   1,0,   1,1,   1,-1,   0,0)}
```

Defining a function that computes a product of CG coefficients

The `comp_CG()` function has the following input parameters:

- T: Dictionary of configuration tuples
- k: Index of the array in the dictionary
- *display*: None by default, set to `True` to display the computation

It returns the following product of CG coefficients pertaining to the entry k:

- $CG(l_1, m_1, l_2, m_2, l_{12}, m_{12}) * CG(l_{12}, m_{12}, l_3, m_3, l_{123}(L), m_{123}(M))$

```
def comp_CG(T, k, display = None):
  CGk = CG(*T[k][0:6]) * CG(*T[k][4:10])
  if display:
    print('CG(', *T[k][0:6], ') = ', CG(*T[k][0:6]).doit())
    print('CG(', *T[k][4:10], ') = ', CG(*T[k][4:10]).doit())
    print("CG{} =".format(k), 'CG(', *T[k][0:6], ') * CG(',
*T[k][4:10], ') = ', CGk.doit())
  return CGk
```

For instance, for $T = T00$ and $k = 0$ with the display option set to `True`, use the following:

```
CG0 = comp_CG(T00, 0, display=True)
```

We get the following detailed output:

```
CG( 1 -1 1 0 1 -1 ) =  -sqrt(2)/2
CG( 1 -1 1 1 0 0 ) =  sqrt(3)/3
CG0 = CG( 1 -1 1 0 1 -1 ) * CG( 1 -1 1 1 0 0 ) =  -sqrt(6)/6
```

Figure 2.10 – Output of comp_CG for the first entry in the T00 dictionary

Computing and printing the CG coefficients

The following Python code calls the comp_CG() function for each entry in the T00 dictionary and prints the result of the computation of the CG coefficients:

```
for k in range(0, len(T00)):
    s = 'CG' + str(k) +' = comp_CG(T00, ' + str(k) + ')'
    exec(s)
s00 = ["CG0: {}, CG1: {}, CG2: {}, CG3: {}, CG4: {}, CG5: {}".
    format(CG0.doit(), CG1.doit(), CG2.doit(), CG3.doit(),
CG4.doit(), CG5.doit())]
print(s00)
```

Here's the result:

```
['CG0: -sqrt(6)/6, CG1: sqrt(6)/6, CG2: -sqrt(6)/6, CG3: sqrt(6)/6, CG4: -sqrt(6)/6, CG5: sqrt(6)/6']
```

Figure 2.11 – CG coefficients for computing the ground state of the nitrogen atom with three p electrons $(L = 0, M = 0)$

Defining a set of spatial wave functions

Since electrons in the same orbital repel one another, we define a set of spatial wave functions, adding a phase of $\pi/3$ and $2\,\pi/3$ in the wave functions of the second and third electron respectively:

```
def Y_phase(theta, phi):
    Y10a = comb_Y(1, 0, theta, phi)
    Y11a = comb_Y(1, 1, theta, phi)
    Y1m1a = comb_Y(1, -1, theta, phi)
    Y10b = comb_Y(1, 0, theta, phi+1*np.pi/3)
    Y11b = comb_Y(1, 1, theta, phi+1*np.pi/3)
    Y1m1b = comb_Y(1, -1, theta, phi+1*np.pi/3)
    Y10c = comb_Y(1, 0, theta, phi+2*np.pi/3)
    Y11c = comb_Y(1, 1, theta, phi+2*np.pi/3)
```

```
    Y1m1c = comb_Y(1, -1, theta, phi+2*np.pi/3)
    return(Y10a, Y11a, Y1m1a, Y10b, Y11b, Y1m1b, Y10c, Y11c,
Y1m1c)
```

Computing the wave function of the Nitrogen atom with three p electrons ($L = 0$, $M = 0$)

We compute the wave function as a sum of the products of the wave functions defined previously:

```
def compute_00_Y(ax_lim, cmap, title,  fig_name):
    fig = plt.figure(figsize=plt.figaspect(1.))
    (theta, phi, xyz) = setup_grid()
    ax = fig.add_subplot(projection='3d')
    (Y10a, Y11a, Y1m1a, Y10b, Y11b, Y1m1b, Y10c, Y11c, Y1m1c) =
Y_phase(theta, phi)
    Y_00 = float(CG0.doit()) * Y1m1a * Y10b * Y11c
    Y_01 = float(CG1.doit()) * Y1m1a * Y11b * Y10c
    Y_02 = float(CG2.doit()) * Y10a * Y1m1b * Y11c
    Y_03 = float(CG3.doit()) * Y10a * Y11b * Y1m1c
    Y_04 = float(CG4.doit()) * Y11a * Y1m1b * Y10c
    Y_05 = float(CG5.doit()) * Y11a * Y10b * Y1m1c
    Y = Y_00 + Y_01 + Y_02 + Y_03 + Y_04 + Y_05
    Yx, Yy, Yz = np.abs(Y) * xyz
    colour_plot(ax, Y, Yx, Yy, Yz, cmap)
    draw_axes(ax, ax_lim, title)
    plt.savefig(fig_name)
    plt.show()
    return
```

Displaying the wave function of the ground state of the nitrogen atom with three p electrons ($L = 0$, $M = 0$)

We now show the graphical representation of the spherical harmonic function for the ground state of the nitrogen atom with three p electrons:

```
title = '$Nitrogen\ with\ 3p\ electrons\ (L=0,\ M=0)$'
fig_name ='Nitrogen_3p_L0_M0.png'
compute_00_Y(0.01, 'autumn', title, fig_name)
```

Here's the result:

Figure 2.12 – Spatial wave function of the ground state of the nitrogen atom with three p electrons
$$(L = 0, M = 0)$$

2.1.3. General formulation of the Pauli exclusion principle

Remember that fermions are particles that have half-integer spin ($s = \frac{1}{2}, \frac{3}{2}, \frac{5}{2}, \cdots$) and bosons are particles that have integer spin ($s = 0,1,2, \ldots$). The general formulation of the PEP states the total wave function Ψ_{total} for a quantum system must have certain symmetries for all sets of identical particles, that is, electrons and identical nuclei, both boson and fermions, under the operation of pair particle permutation [Bubin]:

- For fermions, the total wave function must be antisymmetric ($-$) with respect to the exchange of identical pair particles ($\hat{A}_{ij}$):

$$\hat{A}_{ij}\Psi_{total} = -\Psi_{total}$$

  meaning that the spatial part of the wave function is antisymmetric while the spin part is symmetric, or vice versa.

- For bosons, the total wave function must be symmetric ($+$) with respect to the exchange of pair particles ($\hat{S}_{ij}$):

$$\hat{S}_{ij}\Psi_{total} = +\Psi_{total}$$

meaning that both the spatial wave function and spin function are symmetric, or both are antisymmetric.

- For composite systems with both identical fermions and identical bosons, the preceding operations must hold true simultaneously.

In general, the symmetrizer and antisymmetrizer operations combined for a given quantum system are referred to as the projection operator $\left(\hat{Y}\right)$. The total wave function $(\Psi_{total}(\boldsymbol{r}, \boldsymbol{s}, t))$, including the PEP, is then written as:

$$\Psi_{total}(\boldsymbol{r}, \boldsymbol{s}, t) = \hat{Y}\Psi_{total}(\boldsymbol{r}, \boldsymbol{s}, t)$$

For a given quantum system, the projection operator that satisfies the PEP is obtained as a product of the antisymmetrizer and the symmetrizer, $\hat{Y} = \hat{S}\hat{A}$, and strictly in this order, not $\hat{Y} = \hat{A}\hat{S}$. Making this mistake in a calculation will result in incorrect operations.

The projection operator $\left(\hat{Y}\right)$ can be expressed as a linear combination:

$$\hat{Y} = \sum_{i=1}^{n!} a_i \, \hat{P}_i$$

where the index i indicates a particular order of particles in a set of possible orders, P_i is the permutation associated with a particular order, an associated expansion coefficient a_i, and n is the total number of identical particles. This equation is dependent on a factorial (!) relation of permutations, making this a non-deterministic polynomial time hard (NP-hard) computation. Please note that you cannot add and subtract operations, you can only combine like terms. As the system grows larger in the number of identical particles, the complexity increases exponentially, making this an NP-hard calculation.

The process of determining the symmetrizer $(\hat{S}_{ij})$ and antisymmetrizer $(\hat{A}_{ij})$ for the projection operation to apply PEP to a given quantum system is as follows:

- Identify all sets of identical particles n, that is, electrons and nuclei, and fermions and bosons. Please do not confuse this n for the identical number of particles with the principal quantum number n as we are using the same notation.

- Build a partition function for positive integers. Remember we only have a positive count of particles, not a negative count. A partition of a positive integer n is a sequence of positive integers $(p_1, p_2, \ldots, p_l)$ such that $p_1 \geq p_2 \geq \ldots \geq p_l$ and $p_1 + p_2 + \cdots + p_l = n$, where l is the last possible integer for the set.

- Then use the partition to build a Young frame. A Young frame (diagram) is a series of connected boxes organized in rows that are left-aligned and arranged so that every row contains an equal or lower number of boxes than the row above it.

The totally symmetric irreducible representation of a system with n identical bosons is a vertical Young tableau of n boxes. The totally antisymmetric irreducible representation of a system with n identical fermions and total quantum spin (s), is a horizontal Young tableau of n boxes. We calculate the symmetry quantum (p) number as:

$$p = \left| \frac{n}{2} - s \right|$$

The partition function (μ) describes how to build a Young frame. There are two boxes in the first p rows and one box in the remaining $n - 2p$ rows, which we write as follows:

$$\mu = [2^p 1^{n-2p}]$$

Please note that the superscripts are not exponents. The convention for filling the numbers in the boxes is increasing from left to right, and second increasing from top to bottom. Here are some examples of how to put together the Young frame:

- When there are two identical boson particles ($n = 2$) with total spin $s = 0$, the symmetry quantum number is $p = 1$, the partition function is $\mu = [2^1 1^0]$, and the corresponding Young frames is:

$$\boxed{1}\boxed{2}$$

Figure 2.13 – Young frame for the partition function $\mu = [2^1 1^0]$

This Young frame corresponds to a totally symmetric operation.

- When there are two identical boson particles with total spin $s = 2$, the symmetry quantum number is $p = 1$, the partition function is $\mu = [2^1 1^0]$, and the corresponding Young frame is the same as the previous Young frame.

- When there are two identical fermion particles with total spin $s = 0$, the symmetry quantum number is $p = 1$, the partition function is $\mu = [2^1 1^0]$, and the corresponding Young frame is the same as the previous Young frame. We use this state in *Section 2.2.2, Probability amplitude for a hydrogen anion* (H^-).

- When there are two identical fermion particles with total spin $s = 1$, the symmetry quantum number is $p = 0$, the partition function is $\mu = [2^0 1^2]$, and the corresponding Young frame is as follows:

$$\begin{array}{c}\boxed{1}\\\boxed{2}\end{array}$$

Figure 2.14 – Young frame for the partition function $\mu = [2^0 1^2]$

This Young frame corresponds to a totally antisymmetric operation.

- When there are three identical fermion particles ($n = 3$), with the total spin $s = \frac{1}{2}$, that is, two paired electrons and one lone electron, the symmetry quantum number is $p = 1$, the partition function is $\mu = [2^1 1^1]$, and the corresponding Young frame is as follows:

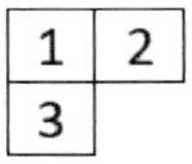

Figure 2.15 – Young frame for the partition function $\mu = [2^1 1^1]$

This Young frame corresponds to both symmetric and antisymmetric operations combined.

- When there are three identical fermion particles ($n = 3$), with the total spin $s = \frac{3}{2}$, that is, three unpaired electrons, the symmetry quantum number is $p = 0$, the partition function is $\mu = [2^0 1^3]$, and the corresponding Young frame is as follows:

Figure 2.16 – Young frame for the partition function $\mu = [2^0 1^3]$

- For the four electrons in lithium hydride (LiH), with spin pairing ($s = 0$), the symmetry quantum number $p = 2$, the partition is $\mu = [2^2 1^0]$, and we have the following Young frame:

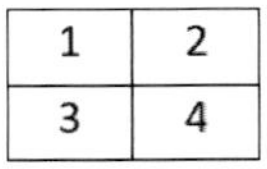

Figure 2.17 – Young frame for the partition function $\mu = [2^2 1^0]$

In this example, since the nucleus is the only particle of its kind, we do not include it in the numbering of the set.

We can generalize the Young frame for fermions, bosons, and composite systems as shown in *Figure 2.18*.

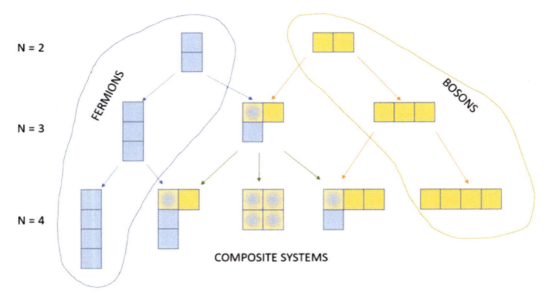

Figure 2.18 – Young frames for fermions, bosons, and composite systems [authors]

The `frame()` function creates a Young frame given a partition as input:

- mu: This partition is represented as a dictionary whose keys are the partition integers and the values are the multiplicity of that integer. For example, $[2^1 1^0]$ is represented as $\{2: 1, 1:0\}$.

It returns a Young frame as follows:

- f: A dictionary of lists whose keys are the index of the lines starting from 0 and the values are the list of integers in the corresponding line. For example, $\{0: [1,2], 1: [3]\}$ represents the Young frame *Figure 2.15* where the first line contains 1,2 and the second line 3:

```
def frame(mu):
    a = 0
    b = 0
    f = {}
    for k, v in mu.items():
        for c in range(v):
            f[a] = list(range(b+1, b+k+1))
```

```
        a += 1
        b += k
    return f
```

Let's run the `frame()` function with $\mu = [2^1 1^0]$:

```
print("F_21_10 =", frame({2: 1, 1:0}))
```

Here is the result:

F_21_10 = {0: [1], 1: [2]}

Let's run the `frame()` function with $\mu = [2^1 1^1]$:

```
print("F_21_11 =", frame{2: 1, 1:1}))
```

Here is the result:

$$F\_21\_11 = \{0: [1,2\ ], 1: [3]\}$$

Now we are ready to define the antisymmetrizer $(\hat{A})$ and symmetrizer $(\hat{S})$ operations for many particles in the system that are governed by the Young frame we determined. The antisymmetrizer operator (A) for the rows of the Young frame is:

$$\hat{A} = \prod^{rows} \sum_{i=1}^{n!} \delta \hat{P}_i$$

where δ is positive for odd permutations and negative for even permutations. An odd permutation has an antisymmetric permutation matrix. An even permutation has a symmetric permutation matrix. We also define a symmetrizer operator $(\hat{S})$ for the columns of the Young frame:

$$\hat{S} = \prod^{columns} \sum_{i=1}^{n!} \delta \hat{P}_i$$

Recall that the projection operator is then the product: $Y = SA$.

For the example of the four electrons in LiH, with spin pairing $(s = 0)$, we derive the following operators from *Figure 2.17*:

$$\hat{A} = \left\{ \left[\mathbb{1} - \hat{P}_{12} \right] \left[\mathbb{1} - \hat{P}_{34} \right] \right\}$$
$$\hat{S} = \left\{ \left[\mathbb{1} + \hat{P}_{13} \right] \left[\mathbb{1} + \hat{P}_{24} \right] \right\}$$

where $\hat{P}_{ab}$: is the permutation of particles a and b, and $\mathbb{1}$ is the Identity operator.

The projection operator is computed using the rules of distributivity and multiplication of permutations:

$$\hat{Y} = \hat{S}\hat{A} = \left[\mathbb{1} + \hat{P}_{13} + \hat{P}_{24} + \hat{P}_{13}\hat{P}_{24}\right]\left[\mathbb{1} - \hat{P}_{12} - \hat{P}_{34} + \hat{P}_{12}\hat{P}_{34}\right]$$
$$\hat{Y} = \hat{S}\hat{A} =$$
$$\mathbb{1} - \hat{P}_{12} - \hat{P}_{34} + \hat{P}_{12}\hat{P}_{34}$$
$$+\hat{P}_{13} - \hat{P}_{13}\hat{P}_{12} - \hat{P}_{13}\hat{P}_{34} + \hat{P}_{13}\hat{P}_{12}\hat{P}_{34}$$
$$+\hat{P}_{24} - \hat{P}_{24}\hat{P}_{12} - \hat{P}_{24}\hat{P}_{34} + \hat{P}_{24}\hat{P}_{12}\hat{P}_{34}$$
$$+\hat{P}_{13}\hat{P}_{24} - \hat{P}_{13}\hat{P}_{24}\hat{P}_{12} - \hat{P}_{13}\hat{P}_{24}\hat{P}_{34} + \hat{P}_{13}\hat{P}_{24}\hat{P}_{12}\hat{P}_{34}$$

With this, we will move on to *Section 2.2, Postulate 2 – Probability amplitude*, where we will revisit the PEP in an example calculation.

2.2. Postulate 2 – Probability amplitude

Consider the motion of a particle in the position space where r is the position vector. The probability density of finding the particle at a particular position and at a given instant in time is calculated as a function of position: $\left|\left|\psi(r)\right\rangle\right|^2$. In an orthonormal basis, the inner product of two wave functions measures their overlap. Two wave functions are orthogonal if their inner product is zero. To find the probability that a state $|\psi(r)\rangle$ will be found in the state $|e_i\rangle$ upon measurement, we must compute the magnitude squared of the inner product between state $|\psi(r)\rangle$ and $|e_i\rangle$, $\theta_i = |\langle\psi(r)|e_i\rangle|^2$.

The wave function in space for a multiparticle system is $|\psi(r_1, \ldots, r_N)\rangle$, with N being the total number of particles, which is interpreted as the probability amplitude function at a given point in time using the following integration over the volume element $d\tau_i$ for all particles in the system $\{1, \ldots, i, \ldots, N\}$:

$$\langle\psi(r_1, \ldots, r_N)|\psi(r_1, \ldots, r_N)\rangle$$
$$= \int \psi(r_1, \ldots, r_N)^\dagger \psi(r_1, \ldots, r_N) dx_1 dy_1 dz_1 \ldots dx_N dy_N dz_N$$
$$= \int_0^\infty \int_0^\pi \int_0^{2\pi} \ldots \int_0^\infty \int_0^\pi \int_0^{2\pi} \psi(r_1, \ldots, r_N)^\dagger \hat{Y}^\dagger \hat{Y}\psi(r_1, \ldots, r_N) r_1^2 \sin\theta_1 dr_1 d\theta_1 d\varphi_1 \ldots r_N^2 \sin\theta_N dr_N d\theta_N d\varphi_N$$

Please note we converted from Cartesian coordinates to spherical coordinates. In this setup, we can include spherical harmonic functions coupled together using CG coefficients that we discussed in the previous section in the wave function $\hat{Y}\psi(r_1, \ldots, r_N)$.

We will also need to include the radial wave functions. We describe how to determine the radial wave functions in *Section 2.2.1, Computing the radial wave functions*, and then go through an example of how to calculate the probability amplitude for a specific quantum chemistry system in *Section 2.2.2, Probability amplitude for a hydrogen anion* (H^-).

2.2.1. Computing the radial wave functions

The radial wave functions for hydrogen-like systems are given by:

$$R(\rho) = \rho^l \sum_{k=0}^{n-l-1} a_k \rho^k e^{-\rho/2}$$

where ρ is calculated by:

$$\rho = \frac{2Z}{na_0} r$$

with a_0 being the Bohr radius set equal to the Committee on Data of the International Science Council (CODATA) value in SI units, and the coefficients a_{k+1} are defined by the following recursion relation:

$$a_{k+1} = a_k(k + l + 1 - n)/((k + 1)(k + 2l + 2))$$

for which the series terminates at $k = n - l - 1$. We initialize a_0 with the following Python code:

```
a0 = 5.29177210903e-11
```

The comp_ak() function has the following input parameters:

- n: Integer, principal quantum number

- l: Angular momentum quantum number with values ranging from 0 to $n - 1$

- a0: Bohr radius, defined by $a_0 = \dfrac{\hbar}{\alpha mc} = 5.29177210903 \times 10^{-11}$, where α is the fine structure constant, c is the speed of light, and m is the rest mass of the electron

- ak: Coefficient defined by the preceding recursion relation

It returns a dictionary whose keys are integers k and values are the corresponding coefficients a_k:

```python
def comp_ak(n):
    n = max(n,1)
    # Create a dictionary with integer keys, starting with a0
    dict = {0: a0}
    for l in range (n-1):
        for k in range(n-l-1):
            ak = dict.get(k)
            #display("l: {}, k: {}, ak: {}".format(l, k, ak))
            dict[k+1] = ak*(k+l+1-n)/((k+1)*(k+2*l+2))
    return dict
```

Let's get the first ten coefficients:

```python
d = comp_ak(10)
for k, v in d.items():
    print("{}, {}".format(k,v))
```

Here is the result:

```
0, 5.29177210903e-11
1, -2.9398733939055554e-12
2, 1.9455044518492647e-13
3, -1.5749321753065475e-14
4, 1.615315051596459e-15
5, -2.2027023430860805e-16
6, 4.2830323337784895e-17
7, -1.3257004842647707e-17
8, 7.954202905588625e-18
9, -1.458270532691248e-17
```

Figure 2.19 – Coefficients that appear in the radial wave functions

Import the SymPy functions:

```python
from sympy.physics.hydrogen import R_nl
from sympy.abc import r, Z
```

The `sympy.physics.hydrogen.Rnl(n,l,r,Z=1)` function returns the hydrogen radial wave function R_{nl} [SymPy_Rnl]. It has the following input parameters:

- n: Integer, principal quantum number

- l: Angular momentum quantum number with values ranging from 0 to n−1

- r: Radial coordinate

- Z: Atomic number (or nuclear charge: 1 for hydrogen, 2 for helium, and so on)

Let's try it first with $n = 1, l = 0$:

```
R_nl(1, 0, r, Z)
```

Here's is the result:

$$2\sqrt{Z^3}e^{-Zr}$$

Next with $n = 2, l = 0$:

```
R_nl(2, 0, r, Z)
```

Here's is the result:

$$\frac{\sqrt{2}}{4}(-Zr + 2)\sqrt{Z^3}e^{-Zr/2}$$

Last with $n = 2, l = 1$:

```
R_nl(2, 1, r, Z)
```

Here's is the result:

$$\frac{\sqrt{6}}{12}Zr\sqrt{Z^3}e^{-Zr/2}$$

2.2.2. Probability amplitude for a hydrogen anion (H⁻)

Let's calculate the probability amplitude at time $t = t_0$ for a hydrogen anion, also called hydride, with one proton and two electrons in a spin paired ground state. This example is for illustration purposes only, and is not meant to be a rigorous calculation.

We label the two electrons as particles 1 and 2 and choose the state where the electronic angular momentum for each electron is $l_1 = 0, m_{l_1} = 0, l_2 = 0, m_{l_2} = 0$, and are coupled to the final or total momenta state of $J = 0, M = 0$, where J is the coupling between the angular momentum and the spin momentum. For simplicity, we assume that this system is not entangled.

We will denote the wave function with the PEP operation $(\hat{Y})$ as:

$$\psi(r_1, r_2, s_1, s_2, t = t_0; J = 0, M = 0, \hat{Y})$$

where the spatial function is symmetric, and the spin function is antisymmetric:

$$\psi(r_1, r_2, s_1, s_2, t = t_0; J = 0, M = 0, \hat{Y}) = CB1 \times |l_{12} = 0, m_{l_{12}} = 0\rangle |s_{12} = 0, s_{l_{12}} = 0\rangle$$

with $CB1$, the CG coefficient, equal to:

$$CB1 = \langle l_{12} = 0, m_{l_{12}} = 0, s_{12} = 0, m_{s_{12}} = 0; j = 0, m_j = 0 | l_{12} = 0, m_{l_{12}} = 0, s_{12} = 0, m_{s_{12}} = 0\rangle = 1$$

Recall that we derived the antisymmetric spin state $|s_{12} = 0, s_{l_{12}} = 0\rangle$ in *Section 2.1.2, Fermionic spin pairing to symmetric state* ($s_{1,2} = 0, m_{s_{1,2}} = 0$), therefore we won't redo this calculation; we will simply reuse the result:

$$\psi(r_1, r_2, s_1, s_2, t = t_0; J = 0, M = 0, \hat{Y}) = \left(\sqrt{2}/2\right) |l_{12} = 0, m_{l_{12}} = 0\rangle \{\uparrow\downarrow - \downarrow\uparrow\}$$

Next, we illustrate the coupling of the angular momentum spatial function for the symmetric spatial state $|l_{12} = 0, m_{l_{12}} = 0\rangle$:

$$|l_{12} = 0, m_{l_{12}} = 0\rangle = CB2 \times |l_1 = 0, m_{l_1} = 0\rangle |l_2 = 0, m_{l_2} = 0\rangle$$

with the CG coefficient $CB2$ equal to:

$$CB2 = \langle l_1 = 0, m_{l_1} = 0, l_2 = 0, m_{l_2} = 0; l_{12} = 0, m_{l_{12}} = 0 | l_1 = 0, m_{l_1} = 0, l_2 = 0, m_{l_2} = 0\rangle = 1$$

Now we plug this into the wave function:

$$\psi(r_1, r_2, s_1, s_2, t = t_0; J = 0, M = 0, \hat{Y}) =$$
$$\left(\sqrt{2}/2\right) |l_1 = 0, m_{l_1} = 0\rangle |l_2 = 0, m_{l_2} = 0\rangle \times \{\uparrow\downarrow - \downarrow\uparrow\}$$

Next, we will be using the following spherical harmonic functions:

$$|l_1 = 0, m_{l_1} = 0\rangle = Y_0^0(\theta_1, \varphi_1) = \frac{1}{2}\sqrt{\frac{1}{\pi}}$$

$$|l_2 = 0, m_{l_2} = 0\rangle = Y_0^0(\theta_2, \varphi_2) = \frac{1}{2}\sqrt{\frac{1}{\pi}}$$

And the radial wave function for each electron with the nuclear charge for the proton of $Z = 1$, as determined in *Section 2.2.1, Computing the radial wave functions*:

$$R(n_1 = 1, l_1 = 0) = 2e^{-r_1}$$
$$R(n_2 = 1, l_2 = 0) = 2e^{-r_2}$$

The wave function for the ground state of hydride is:

$$\psi(r_1, r_2, s_1, s_2, t = t_0; J = 0, M = 0, \hat{Y}) = \left(\frac{\sqrt{2}}{2\pi}\right)e^{-r_1}e^{-r_2}\{\uparrow\downarrow - \downarrow\uparrow\}$$

The probability amplitude is calculated by determining the square of the wave function:

$$\langle\psi(r_1, r_2, s_1, s_2, t = t_0; J = 0, M = 0)|\hat{Y}^\dagger\hat{Y}|\psi(r_1, r_2, s_1, s_2, t = t_0; J = 0, M = 0)\rangle =$$

$$\left(\frac{1}{2\pi^2}\right)\int_0^{+\infty} r_1^2 e^{-2r_1}\, dr_1 \int_0^{+\infty} r_2^2 e^{-2r_2}\, dr_2 \int_0^\pi \sin\theta_1\, d\theta_1 \int_0^\pi \sin\theta_2\, d\theta_2 \int_0^{2\pi} d\varphi_1 \int_0^{2\pi} d\varphi_2$$

$$\times \int \{\uparrow\downarrow - \downarrow\uparrow\}^2 ds_1 ds_2$$

The integral over spin is equal to 1 due to the fact that the spin functions are normalized, resulting in:

$$\langle\psi(r_1, r_2, s_1, s_2, t = t_0; J = 0, M = 0)|\hat{Y}^\dagger\hat{Y}|\psi(r_1, r_2, s_1, s_2, t = t_0; J = 0, M = 0)\rangle =$$

$$\left(\frac{1}{2\pi^2}\right)\int_0^{+\infty} r_1^2 e^{-2r_1}\, dr_1 \int_0^{+\infty} r_2^2 e^{-2r_2}\, dr_2 \int_0^\pi \sin\theta_1\, d\theta_1 \int_0^\pi \sin\theta_2\, d\theta_2 \int_0^{2\pi} d\varphi_1 \int_0^{2\pi} d\varphi_2$$

Next, we include the PEP, where we calculate $\hat{Y}^\dagger \hat{Y} = \hat{A}^\dagger \hat{S} \hat{A}$. Recall that we derived $\hat{Y}$ for two fermions in an antisymmetric spin state as $\left| s_{12} = 0, s_{l_{12}} = 0 \right\rangle$, as shown in *Figure 2.13*. The operation results in a factor of 2:

$$\langle \psi(\boldsymbol{r}_1, \boldsymbol{r}_2, \boldsymbol{s}_1, \boldsymbol{s}_2, t = t_0; J = 0, M = 0) \left| \left[\mathbb{1} + \hat{P}_{12} \right] \right| \psi(\boldsymbol{r}_1, \boldsymbol{r}_2, \boldsymbol{s}_1, \boldsymbol{s}_2, t = t_0; J = 0, M = 0) \rangle =$$

$$\left(\frac{1}{\pi^2} \right) \int_0^{+\infty} r_1^2 e^{-2r_1}\, dr_1 \int_0^{+\infty} r_2^2 e^{-2r_2}\, dr_2 \int_0^{\pi} \sin\theta_1\, d\theta_1 \int_0^{\pi} \sin\theta_2\, d\theta_2 \int_0^{2\pi} d\varphi_1 \int_0^{2\pi} d\varphi_2$$

The integral over r_1 and r_2 is equal to $^1/_4$, illustrated with the following SymPy code:

```
from sympy import symbols, integrate, exp, oo
x = symbols('x')
integrate(x**2 *exp(-2*x),(x,0,oo)).
```

Here is the result:

$$\frac{1}{4}$$

The integrals over θ_1 and θ_2 are equal to 2, illustrated with the following SymPy code:

```
from sympy import symbols, sin, pi
x = symbols('x')
integrate(sin(x),(x,0,pi))
```

Here is the result:

$$2$$

The integrals over φ_1 and φ_2 are equal to 2π, illustrated with the following SymPy code:

```
integrate(1,(x,0,2*pi))
```

Here is the result:

$$2\pi$$

Combining all the results, the probability amplitude is equal to 1:

$$\langle \psi(\boldsymbol{r}_1, \boldsymbol{r}_2, \boldsymbol{s}_1, \boldsymbol{s}_2, t = t_0; J = 0, M = 0, \hat{Y}) | \psi(\boldsymbol{r}_1, \boldsymbol{r}_2, \boldsymbol{s}_1, \boldsymbol{s}_2, t = t_0; J = 0, M = 0, \hat{Y}) \rangle = 1$$

Now we can move on to the rest of the postulates. Examples of these postulates will be illustrated in the following chapters of the book. As a result, we have not included code for these postulates in this chapter. We revisit this topic expectation value in *Section 3.1.9, Pauli matrices*.

2.3. Postulate 3 – Measurable quantities and operators

A physically observable quantity of a quantum system is represented by a linear Hermitian operator, which implies that a measurement outcome is always a real value, not a complex number. The real values of the measurement are the eigenvalues of the Hermitian operator that describes it. The **eigenvalue** is the constant factor that is produced by an operation.

For a spectrum of an observable, if it's discrete, the possible results are quantized. We determine the measurable quantity by calculating the expectation value of the observable $\hat{O}$ in a state $|\psi\rangle$ as follows:

$$\langle \hat{O} \rangle_\psi = \langle \psi | \hat{O} | \psi \rangle = \int_{-\infty}^{\infty} \psi^*(x) \hat{O} \, \psi(x) \, dx$$

It is the sum of all the possible outcomes of a measurement of a state $|\psi\rangle$ weighted by their probabilities. Furthermore, the state of a quantum mechanical system can be represented by the inner product of a given distance called a Hilbert space. A definition of a Hilbert space is given in *Appendix A – Readying Mathematical Concepts*. This definition of a state space implies the **superposition** principle of quantum mechanics, which is a linear combination of all real or complex basis functions ($|\psi\rangle_k$):

$$|\psi\rangle = \sum_{k=1}^{K} c_k \, |\psi\rangle_k$$

where k is the index of summation, K is the total number of basis functions to obtain convergence and completeness of the wave function, and c_k is the linear expansion coefficient, which can be real or complex numbers. Plugging the superposition principle into the definition of the expectation value, we obtain the following equation:

$$\langle \hat{O} \rangle_\psi = \langle \psi | \hat{O} | \psi \rangle = \sum_{k=1}^{K} \sum_{k'=1}^{K} c_k^\dagger c_{k'} \int_{-\infty}^{\infty} \psi_k^\dagger \, \hat{Y}^\dagger \hat{O} \hat{Y} \psi_{k'} \, d\tau$$

where we have also included the PEP. We will use the superposition principle in subsequent chapters. In this section, we present common operators and calculate the expectation value for a given system:

- *Section 2.3.1, Hermitian operator*
- *Section 2.3.2, Unitary operator*
- *Section 2.3.3, Density matrix and mixed quantum states*
- *Section 2.3.4, Position operation with the position operators* $(\hat{r}_{i_x}, \hat{r}_{i_y}, \hat{r}_{i_z})$
- *Section 2.3.5, Momentum operation with the momentum operators* $(\hat{p}_x, \hat{p}_y, \hat{p}_z)$
- *Section 2.3.6, Kinetic energy operation with the kinetic energy operators* $(\hat{T}_x, \hat{T}_y, \hat{T}_z)$
- *Section 2.3.7, Potential energy operation with the potential energy operators* $(\hat{V}_x, \hat{V}_y, \hat{V}_z)$
- *Section 2.3.8, Total energy operation with total energy operators* $(\hat{E}_x, \hat{E}_y, \hat{E}_z)$

The measurable quantum quantities are derived from the classical counterparts.

2.3.1. Hermitian operator

The complex conjugate transpose of some vector a or matrix A often is denoted as $a^\dagger$ and $A^\dagger$ in quantum mechanics. The symbol $\dagger$ is called the dagger. $A^\dagger$ is called the adjoint or Hermitian conjugate of A.

A linear operator U is called Hermitian or self-adjoint if it is its own adjoint: $U^\dagger = U$.

The spectral theorem says that if U is Hermitian then it must have a set of orthonormal eigenvectors:

$$\{|e_i\rangle \,; i \in [1, N], \qquad \langle e_i | e_j \rangle = \delta_{ji}\}$$

where $\delta_{ji} = \begin{cases} 0, & i \neq j \\ 1, & i = j \end{cases}$ with real eigenvalues λ_i, $U|e_i\rangle = \lambda_i |e_i\rangle$, and N is the number of eigenvectors, and also is the dimension of the Hilbert space. Hermitian operators have a unique spectral representation in terms of the set of eigenvalues $\{\lambda_i\}$ and the corresponding eigenvectors $|e_i\rangle$:

$$U = \sum_{i=1}^{N} \lambda_i |e_i\rangle\langle e_i|$$

We revisit this topic in *Section 2.3.3, Density matrix and mixed quantum states*.

Writing matrices as a sum of outer products

The outer product of a ket $|x\rangle$ and a bra $\langle y|$ is the rank-one operator $|x\rangle\langle y|$ with the rule:

$$(|x\rangle\langle y|)(z) = \langle y|z\rangle|x\rangle$$

The outer product of a ket $|x\rangle$ and a bra $\langle y|$ is a simple matrix multiplication:

$$|x\rangle\langle y| \overset{\text{def}}{=} \begin{pmatrix} x_1 \\ x_2 \\ \dots \\ x_n \end{pmatrix} (y_1^*, \quad y_2^*, \quad \dots \quad y_n^*) = \begin{pmatrix} x_1 y_1^* & x_1 y_2^* & \dots & x_1 y_n^* \\ x_2 y_1^* & x_2 y_2^* & \dots & x_2 y_n^* \\ \dots & \dots & \dots & \dots \\ x_n y_1^* & x_n y_2^* & \dots & x_n y_n^* \end{pmatrix}$$

Any matrix can be written in terms of outer products. For instance, for a 2 x 2 matrix:

$$|0\rangle\langle 0| = \begin{pmatrix} 1 \\ 0 \end{pmatrix} (1 \quad 0) = \begin{pmatrix} 1 & 0 \\ 0 & 0 \end{pmatrix} \quad |1\rangle\langle 1| = \begin{pmatrix} 0 \\ 1 \end{pmatrix} (0 \quad 1) = \begin{pmatrix} 0 & 0 \\ 0 & 1 \end{pmatrix}$$

$$|0\rangle\langle 1| = \begin{pmatrix} 1 \\ 0 \end{pmatrix} (0 \quad 1) = \begin{pmatrix} 0 & 1 \\ 0 & 0 \end{pmatrix} \quad |1\rangle\langle 0| = \begin{pmatrix} 0 \\ 1 \end{pmatrix} (1 \quad 0) = \begin{pmatrix} 0 & 0 \\ 1 & 0 \end{pmatrix}$$

$$M = \begin{pmatrix} m_{0,0} & m_{0,1} \\ m_{1,0} & m_{1,1} \end{pmatrix} = m_{0,0}|0\rangle\langle 0| + m_{0,1}|0\rangle\langle 1| + m_{1,0}|1\rangle\langle 0| + m_{1,1}|1\rangle\langle 1|$$

We will be using these matrices in *Chapter 3, Quantum Circuit Model of Computation, Section 3.1.6, Pauli matrices.*

2.3.2. Unitary operator

A linear operator U is called unitary if its adjoint exists and satisfies $U^\dagger U = UU^\dagger = \mathbb{1}$, where $\mathbb{1}$ is the identity matrix, which by definition leaves any vector it is multiplied by unchanged.

Unitary operators preserve inner products:

$$\langle Ux|Uy\rangle = \langle x|U^\dagger U|y\rangle = \langle x|\mathbb{1}|y\rangle = \langle x|y\rangle$$

Hence unitary operators also preserve the norm commonly known as the length of quantum states:

$$\|Ux\| = \langle Ux|Ux\rangle^{\frac{1}{2}} = \langle x|x\rangle^{\frac{1}{2}} = \|x\|$$

For any unitary matrix U, any eigenvectors $|x\rangle$ and $|y\rangle$ and their eigenvalues λ_x and λ_y, $U|x\rangle = \lambda_x|x\rangle$ and $U|y\rangle = \lambda_y|y\rangle$, the eigenvalues λ_x and λ_y have the form $e^{i\theta}$ and if $\lambda_x \neq \lambda_y$ then the eigenvectors $|x\rangle$ and $|y\rangle$ are orthogonal: $\langle x|y\rangle = 0$.

It is useful to note that since for any θ, $\left|e^{i\theta}\right| = 1$:

$$\left|e^{ia} + e^{ib}\right| = \left|e^{\frac{i(a+b)}{2}}\left(e^{\frac{i(a-b)}{2}} + e^{-\frac{i(a-b)}{2}}\right)\right| = \left|e^{\frac{i(a+b)}{2}}\right|\left|e^{\frac{i(a-b)}{2}} + e^{-\frac{i(a-b)}{2}}\right| = \left|e^{\frac{i(a-b)}{2}} + e^{-\frac{i(a-b)}{2}}\right|$$

We will revisit this in *Chapter 3, Quantum Circuit Model of Computation*.

2.3.3. Density matrix and mixed quantum states

Any quantum state, either **mixed** or **pure**, can be described by a **density matrix** (P), which is a normalized positive Hermitian operator where $\rho = \rho^\dagger$. According to the spectral theorem, there exists an orthonormal basis, defined in *Section 2.3.1, Hermitian operator*, such that the density is the sum of all eigenvalues (N):

$$\rho = \sum_{i=1}^{N} \lambda_i |e_i\rangle\langle e_i|$$

where i ranges from 1 to N, λ_i are positive or null eigenvalues ($\lambda_i \geq 0$), and the sum of eigenvalues is the trace operation (tr) of the density matrix and is equal to 1:

$$tr(\rho) = \sum_{i=1}^{N} \lambda_i = 1$$

For example, when the density is $\rho = \begin{pmatrix} \rho_{0,0} & \rho_{0,1} \\ \rho_{1,0} & \rho_{1,1} \end{pmatrix}$, with $\rho = \rho^\dagger$, the trace of the density is:

$$tr(\rho) = \rho_{0,0} + \rho_{1,1} = 1$$

Here are some examples of the density matrices of pure quantum states:

$$\begin{pmatrix} 1 & 0 \\ 0 & 0 \end{pmatrix} = \begin{pmatrix} 1 \\ 0 \end{pmatrix} (1 \quad 0) = |0\rangle\langle 0|$$

$$\begin{pmatrix} 0 & 0 \\ 0 & 1 \end{pmatrix} = \begin{pmatrix} 0 \\ 1 \end{pmatrix} (0 \quad 1) = |1\rangle\langle 1|$$

$$\frac{1}{2}\begin{pmatrix} 1 & -1 \\ -1 & 1 \end{pmatrix} = \frac{1}{2}(|0\rangle - |1\rangle)(|0\rangle - |1\rangle)$$

The density matrix of a mixed quantum state consisting of a statistical ensemble of n pure quantum states $\{|x_i\rangle\,; i \in [1, n]\}$, each with a classical probability of occurrence p_i, is defined as:

$$\rho = \sum_{i=1}^{n} p_i |x_i\rangle\langle x_i|$$

where every p_i is positive or null and their sum is equal to one:

$$tr(\rho) = \sum_{i=1}^{n} p_i = 1$$

We summarize the difference between pure states and mixed states in *Figure 2.20*.

Density matrix of a pure state	Density matrix of a mixed state
$\|\mathbf{x}\rangle$	$\{\|x_i\rangle\,; i \in [1, n]\}$
$\rho = \|x\rangle\langle x\|$	$\rho = \sum_{i=1}^{n} p_i \|x_i\rangle\langle x_i\|$
$tr(\rho^2) = \sum_{i=1}^{N} \lambda_i^2 = 1$	$tr(\rho^2) = \sum_{i=1}^{N} \lambda_i^2 < 1$

Figure 2.20 – Density matrix of pure and mixed quantum states

2.3.4. Position operation

The position observable of particle (j) has the following operators for all directions in Cartesian coordinates:

$$\hat{r}_{j_x}\, \psi(\mathbf{r}_1, \ldots, \mathbf{r}_N) = x_j\, \psi(\mathbf{r}_1, \ldots, \mathbf{r}_N)$$
$$\hat{r}_{j_y}\, \psi(\mathbf{r}_1, \ldots, \mathbf{r}_N) = y_j\, \psi(\mathbf{r}_1, \ldots, \mathbf{r}_N)$$
$$\hat{r}_{j_z}\, \psi(\mathbf{r}_1, \ldots, \mathbf{r}_N) = z_j\, \psi(\mathbf{r}_1, \ldots, \mathbf{r}_N)$$

In spherical coordinates the operations become:

$$\hat{r}_{j_x} \psi(r_1, \dots, r_N) = r_j \sin \theta_j \cos \varphi_j \psi(r_1, \dots, r_N)$$

$$\hat{r}_{j_y} \psi(r_1, \dots, r_N) = r_j \sin \theta_j \sin \varphi_j \psi(r_1, \dots, r_N)$$

$$\hat{r}_{j_z} \psi(r_1, \dots, r_N) = r_j \cos \theta_j \psi(r_1, \dots, r_N)$$

We can calculate the expectation value of the position for a given particle (j) in a chosen direction with the following equation:

$$\langle \psi(r_1, \dots, r_N) | \hat{r}_j | \psi(r_1, \dots, r_N) \rangle = \int \psi(r_1, \dots, r_N)^\dagger \hat{r}_j \psi(r_1, \dots, r_N) dx_1 dy_1 dz_1 \dots dx_N dy_N dz_N$$

For example, using the same system as presented in *Section 2.2.2, Probability amplitude for a hydrogen anion* (H^-), the expectation value of the z-position of electron 1 is determined by:

$$\langle \psi(r_1, r_2; J = 0, M = 0, \hat{Y}) | \hat{r}_{1_x} | \psi(r_1, r_2; J = 0, M = 0, \hat{Y}) \rangle =$$

$$\left(\frac{1}{\pi^2} \right) \int_0^{+\infty} r_1^3 e^{-2r_1} dr_1 \int_0^{+\infty} r_2^2 e^{-2r_2} dr_2 \int_0^{\pi} \sin^2 \theta_1 d\theta_1 \int_0^{\pi} \sin \theta_2 d\theta_2 \int_0^{2\pi} \cos \varphi_1 d\varphi_1 \int_0^{2\pi} d\varphi_2$$

Please note that the integration over r_1 is a cubic function as opposed to a quadratic function, the integration over θ_1 has an additional $\sin \theta_1$, and the integration over φ_1 has a $\cos \varphi_1$ as compared to what is seen in the *Section 2.2.2, Probability amplitude for a hydrogen anion* (H^-) example. In this calculation, the integration over φ_1 is equal to 0, which means that the entire integration is:

$$\langle \psi(r_1, r_2; J = 0, M = 0, \hat{Y}) | \hat{r}_{1_x} | \psi(r_1, r_2; J = 0, M = 0, \hat{Y}) \rangle = 0$$

This means that electron 1 is most likely to be found at the nucleus (or the origin of the coordinate system). The same holds true for $\hat{r}_{1_y}, \hat{r}_{1_z}, \hat{r}_{2_x}, \hat{r}_{2_y},$ and $\hat{r}_{2_z}$ operations.

2.3.5. Momentum operation

The component of momentum operator for particle (j) is $\hat{p}_{j_x}$ along the x-dimension (and similarly, for the y- and z-dimensions) and is defined as follows in Cartesian coordinates:

$$\hat{p}_{j_x} = -i\hbar \frac{\partial}{\partial x_j}$$

$$\hat{p}_{j_y} = -i\hbar \frac{\partial}{\partial y_j}$$

$$\hat{p}_{j_z} = -i\hbar \frac{\partial}{\partial z_j}$$

We can also write these operators in terms of the spherical derivatives [ucsd]:

$$\hat{p}_{j_x} = -i\hbar \left(\sin\theta_j \cos\varphi_j \frac{\partial}{\partial r_j} + \frac{1}{r_j}\cos\theta_j \cos\varphi_j \frac{\partial}{\partial \theta_j} - \frac{1}{r_j}\frac{\sin\varphi_j}{\sin\theta_j}\frac{\partial}{\partial \varphi_j} \right)$$

$$\hat{p}_{j_y} = -i\hbar \left(\sin\theta_j \sin\varphi_j \frac{\partial}{\partial r_j} + \frac{1}{r_j}\cos\theta_j \sin\varphi_j \frac{\partial}{\partial \theta_j} - \frac{1}{r_j}\frac{\cos\varphi_j}{\sin\theta_j}\frac{\partial}{\partial \varphi_j} \right)$$

$$\hat{p}_{j_z} = -i\hbar \left(\cos\theta_j \frac{\partial}{\partial r_j} - \frac{1}{r_j}\sin\theta_j \frac{\partial}{\partial \theta_j} \right)$$

We can calculate the expectation value of the momentum for a given particle (j) with the following equation:

$$\langle \psi(r_1, \dots, r_N) | \hat{p}_{j_w} | \psi(r_1, \dots, r_N) \rangle = \int \psi(r_1, \dots, r_N)^\dagger \, \hat{p}_{j_w} \, \psi(r_1, \dots, r_N) dx_1 dy_1 dz_1 \dots dx_N dy_N dz_N$$

where we use w as a generic dimension.

For example, using the same system as presented in *Section 2.2.2, Probability amplitude for a hydrogen anion* (H$^-$), the derivative for the z-momentum operator of electron 1 is:

$$\hat{p}_{1_z}\psi\left(r_1, r_2, s_1, s_2, t = t_0; L = 0, M = 0, \hat{Y}\right) = -i\hbar \left(\frac{\sqrt{2}}{2\pi}\right) \cos\theta_1 \left(\frac{\partial}{\partial r_1}e^{-r_1}\right)e^{-r_2}\{\uparrow\downarrow - \downarrow\uparrow\}$$

where the derivative is:

$$\frac{\partial}{\partial r_1}e^{-r_1} = -e^{-r_1}$$

Therefore, the expectation value of the z-momentum for electron 1 is:

$$\langle \psi(\boldsymbol{r}_1, \boldsymbol{r}_2; J = 0, M = 0, \hat{Y}) | \hat{p}_{1_z} | \psi(\boldsymbol{r}_1, \boldsymbol{r}_2; J = 0, M = 0, \hat{Y}) \rangle =$$

$$i\hbar \left(\frac{1}{\pi^2} \right) \int_0^{+\infty} r_1^2 e^{-2r_1} \, dr_1 \int_0^{+\infty} r_2^2 e^{-2r_2} \, dr_2 \int_0^\pi \cos\theta_1 \sin\theta_1 \, d\theta_1 \int_0^\pi \sin\theta_2 \, d\theta_2 \int_0^{2\pi} d\varphi_1 \int_0^{2\pi} d\varphi_2 = 0$$

which, due to the integration over θ_1, becomes equal to 0, as illustrated by the following SymPy code:

```
from sympy import symbols, sin, cos
x = symbols('x')
integrate(cos(x)*sin(x),(x,0,pi))
```

Here is the result:

0

This result is intuitive because we are in a $J = 0, M = 0$ system, which does not have momentum.

2.3.6. Kinetic energy operation

The kinetic energy operators for a single particle in a given direction $\left(\hat{T}_{j_x}, \hat{T}_{j_y}, \hat{T}_{j_z} \right)$ in Cartesian coordinates are:

$$\hat{T}_{j_x} = -\left(\frac{\hbar^2}{2m_j} \right) \frac{\partial^2}{\partial x_j^2}$$

$$\hat{T}_{j_y} = -\left(\frac{\hbar^2}{2m_j} \right) \frac{\partial^2}{\partial y_j^2}$$

$$\hat{T}_{j_z} = -\left(\frac{\hbar^2}{2m_j} \right) \frac{\partial^2}{\partial z_j^2}$$

In general, kinetic energy is determined by the following in Cartesian coordinates:

$$\nabla_j^2 = \frac{\partial^2}{\partial x_j^2} + \frac{\partial^2}{\partial y_j^2} + \frac{\partial^2}{\partial z_j^2}$$

and in spherical coordinates is:

$$\nabla_j^2 = \frac{1}{r_j^2}\frac{\partial}{\partial r_j}\left(r_j^2\frac{\partial}{\partial r_j}\right) + \frac{1}{r_j^2\sin\theta_j}\frac{\partial}{\partial\theta_j}\left(\sin\theta_j\frac{\partial}{\partial\theta_j}\right) + \frac{1}{r_j^2\sin^2\theta_j}\frac{\partial^2}{\partial\varphi_j^2}$$

We can calculate the expectation value of the kinetic energy for all the particles with the following equation:

$$\langle\hat{T}\rangle = -\frac{\hbar^2}{2}\sum_{j=1}^{N}\frac{1}{m_j}\langle\psi(\mathbf{r}_1,\dots,\mathbf{r}_N)|\nabla_j^2|\psi(\mathbf{r}_1,\dots,\mathbf{r}_N)\rangle$$

$$= -\frac{\hbar^2}{2}\sum_{j=1}^{N}\frac{1}{m_j}\int\psi(\mathbf{r}_1,\dots,\mathbf{r}_N)^{\dagger}\,\nabla_j^2\psi(\mathbf{r}_1,\dots,\mathbf{r}_N)dx_1dy_1dz_1\dots dx_Ndy_Ndz_N$$

Using the same system as presented in *Section 2.2.2, Probability amplitude for a hydrogen anion* (H^-), the second derivative operation for the kinetic energy of electron 1 is:

$$\nabla_1^2\psi(\mathbf{r}_1,\mathbf{r}_2,\mathbf{s}_1,\mathbf{s}_2,t=t_0;L=0,M=0,\hat{Y}) = \left(\frac{\sqrt{2}}{2\pi}\right)e^{-r_2}\{\uparrow\downarrow-\downarrow\uparrow\}\frac{1}{r_1^2}\frac{\partial}{\partial r_1}\left(r_1^2\frac{\partial}{\partial r_1}\right)(e^{-r_1})$$

$$= -\left(\frac{\sqrt{2}}{2\pi}\right)e^{-r_2}\{\uparrow\downarrow-\downarrow\uparrow\}\frac{1}{r_1^2}\frac{\partial}{\partial r_1}(r_1^2e^{-r_1})$$

$$= \left(\frac{\sqrt{2}}{2\pi}\right)e^{-r_2}\{\uparrow\downarrow-\downarrow\uparrow\}\frac{1}{r_1^2}(r_1^2-2r_1)e^{-r_1}$$

$$= \left(\frac{\sqrt{2}}{2\pi}\right)\left(1-\frac{2}{r_1}\right)e^{-r_1}e^{-r_2}\{\uparrow\downarrow-\downarrow\uparrow\}$$

The expectation value of the kinetic energy for electron 1 is then calculated by:

$$-\left(\frac{\hbar^2}{2m_1}\right)\langle\psi(\mathbf{r}_1,\mathbf{r}_2,\mathbf{s}_1,\mathbf{s}_2;J=0,M=0,\hat{Y})|\nabla_1^2|\psi(\mathbf{r}_1,\mathbf{r}_2,\mathbf{s}_1,\mathbf{s}_2;J=0,M=0,\hat{Y})\rangle =$$

$$-\left(\frac{\hbar^2}{m_1}\right)\left(\frac{1}{\pi^2}\right)\int_0^{+\infty}r_1^2e^{-2r_1}\,dr_1\int_0^{+\infty}r_2^2e^{-2r_2}\,dr_2\int_0^{\pi}\sin\theta_1\,d\theta_1\int_0^{\pi}\sin\theta_2\,d\theta_2\int_0^{2\pi}d\varphi_1\int_0^{2\pi}d\varphi_2$$

$$+2\left(\frac{\hbar^2}{m_1}\right)\left(\frac{1}{\pi^2}\right)\int_0^{+\infty}r_1e^{-2r_1}\,dr_1\int_0^{+\infty}r_2^2e^{-2r_2}\,dr_2\int_0^{\pi}\sin\theta_1\,d\theta_1\int_0^{\pi}\sin\theta_2\,d\theta_2\int_0^{2\pi}d\varphi_1\int_0^{2\pi}d\varphi_2$$

$$= \left(\frac{\hbar^2}{m_1}\right)\{-1+2\} = \hbar^2$$

where the electron mass is set to equal 1 ($m_1 = 1$). The kinetic energy for electron 2 is determined with the same integrals and is equal to:

$$-\left(\frac{\hbar^2}{2m_2}\right)\langle\psi(\boldsymbol{r}_1,\boldsymbol{r}_2,\boldsymbol{s}_1,\boldsymbol{s}_2;J=0,M=0,\hat{Y})|\nabla_2^2|\,\psi(\boldsymbol{r}_1,\boldsymbol{r}_2,\boldsymbol{s}_1,\boldsymbol{s}_2;J=0,M=0,\hat{Y})\rangle = \hbar^2$$

The total kinetic energy for the electrons in hydride is then the sum of the two kinetic terms:

$$\langle\hat{T}\rangle = \langle\hat{T}_1 + \hat{T}_2\rangle = 2\hbar^2$$

Setting $\hbar = 1$ as a standard scaling, we have:

$$\langle\hat{T}\rangle = 2$$

2.3.7. Potential energy operation

The potential energy, also known as Coulomb energy, relates the charge $\left(Q_i, Q_j\right)$ of particles i and j and depends on the distance r_{ij} between two, where $r_{ij} = |\boldsymbol{r}_j - \boldsymbol{r}_i| = \sqrt{(x_j - x_i)^2 + (y_j - y_i)^2 + (z_j - z_i)^2}$. It is proportional to the inverse of the distance $\left(1/r_{ij}\right)$ and is calculated as a sum over all pairs of particles in the systems:

$$\hat{V}\left(r_1, r_i, r_j, \ldots r_N\right) = \sum_{i=1<j}^{N} \frac{Q_i Q_j}{r_{ij}}$$

We can calculate the expectation value of the potential energy for all the particles with the following equation:

$$\langle\hat{V}\rangle = \sum_{i=1,\;j=i+1}^{N-1,\;N} Q_i Q_j \left\langle\psi(\boldsymbol{r}_1, \ldots, \boldsymbol{r}_N)\left|\frac{1}{r_{ij}}\right|\psi(\boldsymbol{r}_1, \ldots, \boldsymbol{r}_N)\right\rangle$$

Using the same system as presented in *Section 2.2.2, Probability amplitude for a hydrogen anion* (H^-), the expectation value of the potential (Coulomb) energy calculated between the two electrons is:

$$\langle \hat{V}_{12} \rangle = Q_1 Q_2 \left\langle \psi(r_1, r_2, s_1, s_2; J = 0, M = 0, \hat{Y}) \left| \frac{[1 + \hat{P}_{12}]}{r_{12}} \right| \psi(r_1, r_2, s_1, s_2; J = 0, M = 0, \hat{Y}) \right\rangle$$

$$= \left(\frac{1}{\pi^2} \right) \int_0^{+\infty} \frac{r_1^2 e^{-2r_1} r_2^2 e^{-2r_2}}{|r_1 - r_2|} dr_1 dr_2 \int_0^{\pi} \sin\theta_1 \, d\theta_1 \int_0^{\pi} \sin\theta_2 \, d\theta_2 \int_0^{2\pi} d\varphi_1 \int_0^{2\pi} d\varphi_2$$

$$= 16 \int_0^{+\infty} \frac{r_1^2 e^{-2r_1} r_2^2 e^{-2r_2}}{|r_1 - r_2|} dr_1 dr_2$$

Now we use the Dirac delta function $\delta(r_1 - r_2)$ to approximate the inverse of r_{12}:

$$\langle \hat{V}_{12} \rangle = 16 \int_0^{+\infty} r_1^2 e^{-2r_1} r_2^2 e^{-2r_2} \delta(r_1 - r_2) \, dr_1 dr_2$$

We compute this integral with the following block of code:

```
from sympy import symbols, integrate, exp, DiracDelta, oo
x, y = symbols('x y')
integrate(x**2 * exp(-2*x) * integrate(y**2 * exp(-
2*y)*DiracDelta(x - y),(y,0,oo)),(x,0,oo))
```

The result is:

$$\frac{3}{128}$$

Therefore, the expectation value of electron repulsion is:

$$\langle \hat{V}_{12} \rangle = 16 \times \frac{3}{128} = \frac{3}{8}$$

The expectation value of the potential (Coulomb) energy calculated between electron 1 and the nucleus (particle 3) is:

$$\langle \hat{V}_{13} \rangle = Q_1 Q_3 \left\langle \psi(\boldsymbol{r}_1, \boldsymbol{r}_2, \boldsymbol{s}_1, \boldsymbol{s}_2; J = 0, M = 0, \hat{Y}) \left| \frac{[1 + \hat{P}_{12}]}{r_{13}} \right| \psi(\boldsymbol{r}_1, \boldsymbol{r}_2, \boldsymbol{s}_1, \boldsymbol{s}_2; J = 0, M = 0, \hat{Y}) \right\rangle =$$

$$-\left(\frac{1}{\pi^2}\right) \int_0^{+\infty} \frac{r_1^2 e^{-2r_1}}{|r_1 - r_3|} dr_1 \int_0^{+\infty} r_2^2 e^{-2r_2} \, dr_2 \int_0^{\pi} \sin\theta_1 \, d\theta_1 \int_0^{\pi} \sin\theta_2 \, d\theta_2 \int_0^{2\pi} d\varphi_1 \int_0^{2\pi} d\varphi_2 =$$

$$-4 \int_0^{+\infty} \frac{r_1^2 e^{-2r_1}}{|r_1 - r_3|} dr_1$$

Now we use the Dirac delta function $\delta(r_1 - r_3)$ to approximate the inverse of r_{13}:

$$\int_0^{+\infty} r_1^2 e^{-2r_1} \, \delta(r_1 - r_3) dr_1$$

We compute this integral with the following block of code:

```
from sympy import symbols, integrate, exp, DiracDelta, oo
x, y = symbols('x y')
integrate(x**2 * exp(-2*x) * integrate(DiracDelta(x -
y),(y,0,oo)),(x,0,oo))
```

Here is the result:

$$\frac{1}{4}$$

Therefore, the expectation value of electron-nuclear attraction is:

$$\langle \hat{V}_{13} \rangle = -4 \times \frac{1}{4} = -1 = \langle \hat{V}_{23} \rangle$$

The total potential energy is:

$$\langle \hat{V} \rangle = \langle \hat{V}_{12} \rangle + \langle \hat{V}_{13} \rangle + \langle \hat{V}_{23} \rangle = \frac{3}{8} - 1 - 1 = -\frac{13}{8}$$

2.3.8. Total energy operation

The total energy operator $\left(\hat{H}\right)$ is the sum of the kinetic energy and the potential energy operations:

$$\hat{H}|\psi\rangle = -\left(\frac{\hbar^2}{2}\right)\sum_{j=1}^{N}\frac{1}{m_j r_j^2}\left[\frac{\partial}{\partial r_j}\left(r_j^2\frac{\partial|\psi\rangle}{\partial r_j}\right) + \frac{1}{\sin\theta_j}\frac{\partial}{\partial\theta_j}\left(\sin\theta_j\frac{\partial|\psi\rangle}{\partial\theta_j}\right) + \frac{1}{\sin^2\theta_j}\frac{\partial^2|\psi\rangle}{\partial\varphi_j^2}\right]$$
$$+ \hat{V}\left(r_1, r_i, r_j, \dots r_N\right)|\psi\rangle = E|\psi\rangle$$

where E is the total energy. The expectation value for the energy is then:

$$\langle\hat{H}\rangle = \langle\hat{T}\rangle + \langle\hat{V}\rangle$$

Using the same system as presented in *Section 2.2.2, Probability amplitude for a hydrogen anion* (H^-), the expectation value of the total energy is:

$$\langle\hat{H}\rangle = 2 - \frac{13}{8} = \frac{3}{8}$$

Notice that the expectation value for hydride is dominated by the potential energy, which makes the system very reactive.

2.4. Postulate 4 – Time-independent stationary states

A quantum state is a time-independent stationary state if all its observables are independent of time. These states are very important in quantum chemistry. The atomic orbital of an electron and the molecular orbital of an electron in a molecule are time-independent stationary states.

The time-independent Schrödinger equation can be written as follows, that is, static: $\hat{H}|\psi\rangle = E|\psi\rangle$ where E is the energy eigenvalue, and $|\psi\rangle$ is the state vector of the quantum system not as a function of time.

This postulate implies that the wave function must be an eigenfunction for all measurements and corresponding operations that represent the energy. An eigenfunction is a function that remains unchanged when acted upon it by an operator or when a measurement is made.

We use this concept more in *Chapter 4, Molecular Hamiltonians*.

2.5. Postulate 5 – Time evolution dynamics

The time evolution dynamics of a quantum system is described by Schrödinger's equation:

$$i\hbar\frac{d}{dt}|\psi\rangle = \widehat{H}|\psi\rangle$$

We will be showing an example of this in *Chapter 5, Variational Quantum Eigensolver (VQE) Algorithm*.

Questions

Please test your understanding of the concepts presented in this chapter with the corresponding Google Colab notebook.

1. What quantum numbers do the total wave function depend on?
2. What is the CG coefficient if we couple together $l_1 = 0$ and $m_1 = 0$ and $l_2 = 1$, $m_2 = 0$ to $L = 1$, $M = 0$?
3. What happens to the total wave function upon the application of an antisymmetric operation?
4. For a pure fermionic state, is the Young frame horizontal or vertical?
5. What is the position operator for the z-direction?
6. What is the sum of potential and kinetic energy?

Answers

1. n, l, m_l and s, m_s
2. 1
3. It is multiplied by -1
4. Vertical
5. $\widehat{r}_{j_z}\,\psi(r_1,\dots,r_N) = z_j\,\psi(r_1,\dots,r_N)$
6. Total energy

References

[Bubin] Bubin, S., Cafiero, M., & Adamowicz, L., Non-Born-Oppenheimer variational calculations of atoms and molecules with explicitly correlated Gaussian basis functions, Advances in Chemical Physics, 131, 377-475, https://doi.org/10.1002/0471739464.ch6

[Cmap] Choosing Colormaps in Matplotlib, https://matplotlib.org/stable/tutorials/colors/colormaps.html

[Lucr_1] Lucretius on the Nature of Things, Literally translated into English prose by the Rev. John Selby Watson, M.A., London 1870, https://www.google.fr/books/edition/Lucretius_On_the_Nature_of_Things/59HTAAAAMAAJ?hl=en&gbpv=1&printsec=frontcover

[Lucr_2] Thomas Nail, Lucretius: Our Contemporary, 15 Feb 2019, https://www.youtube.com/watch?v=VMrTk1A2GX8

[Lucr_3] David Goodhew, Lucretius lecture, Life, love, death and atomic physics, https://www.youtube.com/watch?v=mJZZd3f_-oE

[NumPy] NumPy: the absolute basics for beginners, https://numpy.org/doc/stable/user/absolute_beginners.html

[Phys5250] Addition of angular momentum, University of Colorado, PHYS5250, https://physicscourses.colorado.edu/phys5250/phys5250_fa19/lecture/lec32-addition-angular-momentum/

[SciPy_sph] SciPy, API reference, Compute spherical harmonics, scipy.special.sph_harm, https://docs.scipy.org/doc/scipy/reference/generated/scipy.special.sph_harm.html

[Sharkey_0] Keeper L. Sharkey and Ludwik Adamowicz, An algorithm for nonrelativistic quantum mechanical finite-nuclear-mass variational calculations of nitrogen atom in L = 0, M = 0 states using all-electrons explicitly correlated Gaussian basis functions, J. Chem. Phys. 140, 174112 (2014), https://doi.org/10.1063/1.4873916

[SymPy_CG] SymPy, Clebsch-Gordan Coefficients, https://docs.sympy.org/latest/modules/physics/quantum/cg.html

[SymPy_Rnl] Hydrogen Wavefunctions, https://docs.sympy.org/latest/modules/physics/hydrogen.html

[SymPy] SymPy, A Python library for symbolic mathematics, https://www.sympy.org/en/index.html

[Sph_Real] Wikipedia, Spherical Harmonics, Real forms, `https://en.wikipedia.org/wiki/Spherical_harmonics#Real_forms`

[Ucsd] University of Californian San Diego, Spherical Coordinates and the Angular Momentum Operators, `https://quantummechanics.ucsd.edu/ph130a/130_notes/node216.html`

[Wiki_1] Mathematical formulation of quantum mechanics, Wikipedia, `https://en.wikipedia.org/wiki/Mathematical_formulation_of_quantum_mechanics`

3

Quantum Circuit Model of Computation

"As we scale towards a million [qubits], I think we've got some fundamental issues in error correction, control, and maybe quantum physics that can rear their heads," he said, adding that even those problems are "solvable."

– Arvind Krishna, IBM chairman and CEO

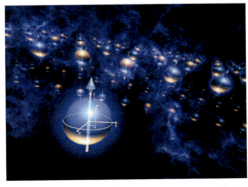

Figure 3.1 – Scaling of the quantum computer [authors]

There are fundamental differences between classical computing and quantum computing; classical computing is deterministic with 1s and 0s, and quantum is probabilistic with a twist. Quantum computers work with probability amplitudes, which is a postulate of quantum mechanics (see *Section 2.2, Postulate 2 – Probability amplitudes*). The probabilistic amplitudes of quantum computing behave differently from classical probabilities in that these values can cancel each other out, which is known as **destructive interference**.

Destructive interference can be illustrated with noise-canceling headphones. Specifically, it is when two or more waves come together, eliminating the waves altogether. In other words, the waves that come together are opposite in phase and equal in amplitude. **Constructive interference** is when two or more waves come together, and the amplitudes add positively. These two properties are essential to enable the desired result to come out of the computer with the highest probability.

So, interference is at the core of what quantum computing should be, and we also use the concept of the wave function as introduced in *Section 2.1, Postulate 1 – Wave function*, and it is used to define the idea of the **qubit**: the quantum bit of information. Typically, in the quantum computing industry, **state vector** is used as the term for the wave function.

Another difference between the two methods of computing is that in quantum computing, when we add one more unit of information, the size of the computational space is doubled. In theory, this allows us to speed up exponentially.

Quantum computing also uses the superposition property to achieve parallelism up until the moment a measurement is performed. Recall we discussed superposition in *Section 2.3, Postulate 3 – Measurable quantities and operators*. A quantum algorithm needs to be repeated multiplied times to get the probability distribution of the measurement.

The scaling of quantum computers paves the way for simulating chemical systems that could enable researchers to conduct virtual experiments and discover new molecules much faster than by performing physical experiments in a lab. In parallel to building a scalable quantum computer, research into optimal mappings of fermionic states and operators to qubit states and quantum gates is essential to exploit the potential of near-term quantum computers.

We give an illustration of a key component of such mappings, a quantum circuit that creates permutation symmetric or permutation asymmetric states in a probabilistic manner.

In this chapter, we will cover the following topics:

- *Section 3.1, Qubits, entanglement, Bloch sphere, Pauli matrices*
- *Section 3.2, Quantum gates*
- *Section 3.3, Computation-driven interference*
- *Section 3.4, Preparing a permutation symmetric or asymmetric state*

Technical requirements

A companion Jupyter notebook for this chapter can be downloaded from GitHub at `https://github.com/PacktPublishing/Quantum-Chemistry-and-Computing-for-the-Curious`, which has been tested in the Google Colab environment, which is free and runs entirely in the cloud, and in the IBM Quantum Lab environment. Please refer to *Appendix B – Leveraging Jupyter Notebooks in the Cloud*, for more information. The companion Jupyter notebook automatically installs the following list of libraries:

- **Numerical Python (NumPy)** [NumPy], an open-source Python library that is used in almost every field of science and engineering

- Qiskit [Qiskit], an open-source SDK for working with quantum computers at the level of pulses, circuits, and application modules

- Qiskit visualization support to enable visualization and Jupyter notebooks

- **Quantum Toolbox in Python (QuTiP)** [QuTiP], which is designed to be a general framework for solving quantum mechanics problems such as systems composed of few-level quantum systems and harmonic oscillators

We recommend using the following online graphical tools:

- IBM Quantum Composer, which is a graphical quantum programming tool that lets you drag and drop operations to build quantum circuits and run them on real quantum hardware or simulators [IBM_comp1] [IBM_comp2]

- Grok the Bloch Sphere, a web-based application that displays the Bloch sphere and shows the action of gates as rotations [Grok]

Installing NumPy, Qiskit, QuTiP, and importing various modules

Install NumPy with the following command:

```
pip install numpy
```

Install Qiskit with the following command:

```
pip install qiskit
```

Install Qiskit visualization support with the following command:

```
pip install 'qiskit[visualization]'
```

Install QuTiP with the following command:

```
pip install qutip
```

Import NumPy with the following command:

```
import numpy as np
```

Import the required functions and class methods. The `array_to_latex function()` returns a LaTeX representation of a complex array with dimension 1 or 2:

```
from qiskit.visualization import array_to_latex, plot_bloch_
vector, plot_bloch_multivector, plot_state_qsphere, plot_state_
city
from qiskit import QuantumRegister, ClassicalRegister,
QuantumCircuit, transpile
from qiskit import execute, Aer
import qiskit.quantum_info as qi
from qiskit.extensions import Initialize
from qiskit.providers.
aer import extensions  # import aer snapshot instructions
```

Import the math libraries with the following commands:

```
import cmath
import math
```

Import QuTiP with the following command:

```
import qutip
```

3.1. Qubits, entanglement, Bloch sphere, Pauli matrices

The concepts presented in this section are a specific application of the five postulates of quantum mechanics that were presented in *Chapter 2, Postulates of Quantum Mechanics*.

In this section, we describe the following in detail:

- *Section 3.1.1, Qubits*
- *Section 3.1.2, Tensor ordering of qubits*

3.1.1. Qubits

In this section, we describe the current setup for quantum computation and the definition of a qubit. A qubit is a unit of information that represents a two-level quantum system and lives in a two-dimensional Hilbert space $\mathbb{C}^2$. The basis vectors of the quantum space are denoted as $\{|0\rangle, |1\rangle\}$, which are referred to as the computational basis states:

$$|0\rangle = \begin{pmatrix} 1 \\ 0 \end{pmatrix} \qquad |1\rangle = \begin{pmatrix} 0 \\ 1 \end{pmatrix}$$

A general single-qubit state is described by a superposition of the computational basis:

$$|\psi\rangle = \alpha|0\rangle + \beta|1\rangle = \begin{pmatrix} \alpha \\ \beta \end{pmatrix} \in \mathbb{C}^2$$

where α and β are linear expansion coefficients that satisfy:

$$|\alpha|^2 + |\beta|^2 = 1$$

Although the qubit is in a quantum superposition during the algorithm, when it is measured in the computational basis, it will be found in state $|0\rangle$ or state $|1\rangle$, not in a superposition. These measurement outcomes occur with probability $|\alpha|^2$ and $|\beta|^2$ respectively. If there are n qubits in the system, the state is described by a vector in the 2^n dimensional Hilbert space $(\mathbb{C}^2)^{\otimes n}$ formed by taking the tensor product of the Hilbert spaces of the individual qubits. For 10 qubits, the state is described by a vector in a 1,024-dimensional Hilbert space.

3.1.2. Tensor ordering of qubits

The physics community typically orders a tensor product of n qubits with the 0th qubit on the left-most side of the tensor product:

$$|q\rangle = |q_0\rangle|q_1\rangle \dots |q_{n-1}\rangle = |q_0, q_1, \dots, q_{n-1}\rangle = \bigotimes_{i=0}^{n-1} |q_i\rangle$$

where $q_i \in \{0,1\}$. However, Qiskit uses an ordering in which the nth qubit is first in the order and the 0th qubit is last:

$$|q\rangle = |q_{n-1}\rangle ... |q_1\rangle|q_0\rangle = |q_{n-1}, ..., q_1, q_0\rangle = \bigotimes_{i=n-1}^{0} |q_i\rangle$$

In other words, if qubit 0 is in state $|0\rangle$, qubit 1 is in state $|0\rangle$, and qubit 2 is in state $|1\rangle$, many physics textbooks would represent this as $|001\rangle$, whereas Qiskit would represent this state as $|100\rangle$. This difference affects the way multi-qubit operations are represented as matrices, so please be on the lookout as we are using Qiskit in the book.

3.1.3. Quantum entanglement

A quantum system is entangled when its quantum state cannot be factored as a tensor product of states of its constituents. States can be classified as either a product of single particle states or **entangled**:

- Product states can be decomposed into tensor products of fewer qubits, such as:

$$\frac{1}{\sqrt{2}}(|00\rangle + |01\rangle) = |0\rangle \otimes \frac{1}{\sqrt{2}}(|0\rangle + |1\rangle)$$

- Entangled states cannot be decomposed into tensor products of states. For example, the Bell state $\frac{1}{\sqrt{2}}(|00\rangle + |11\rangle)$ is entangled and can only be measured either in the state $|00\rangle$ or in the state $|11\rangle$, each with a probability of 1/2.

3.1.4. Bloch sphere

The Bloch sphere describes a qubit in space and is a specific form of the coordinate system (*Figure 3.2*) presented in *Section 2.1, Postulate 1 – Wave functions*. The r vector, or the length, of a qubit is always equal to 1, so the coordinates of the Bloch sphere are:

$$\begin{pmatrix} x \\ y \\ z \end{pmatrix} = \begin{pmatrix} \sin\theta \, \cos\varphi \\ \sin\theta \sin\varphi \\ \cos\theta \end{pmatrix}$$

Let's focus on the general normalized pure state for a single qubit, as presented in *Section 3.1.1, Qubits*:

- When $\alpha = 1$ and $\beta = 0$, the $|0\rangle$ state is "up" in the z-direction.
- When $\alpha = 0$ and $\beta = 1$, the $|1\rangle$ state is "down" in the z-direction.

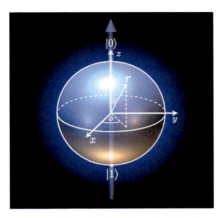

Figure 3.2 – Bloch sphere [authors]

We derived the generalized formula for a qubit in *Section 2.1.2, Addition of momenta using Clebsch-Gordan coefficients, example: Coupling spin and angular momentum* ($j_{1,2} = \frac{1}{2}, js_{1,2} = \frac{1}{2}$):

$$\left| j_{1,2} = \frac{1}{2}, js_{1,2} = \frac{1}{2} \right\rangle = \{\cos\theta \uparrow + e^{-i\varphi}\sin\theta \downarrow\}$$

However, established by the quantum computing industry and set as a convention, we have a change of variables for the angles θ and φ for a qubit defined on the Bloch sphere where the following applies:

- $\theta = 2\arccos(|\alpha|)$ in $[0, \pi]$, which becomes $\frac{\theta}{2} = \arccos(|\alpha|)$ in $\left[0, \frac{\pi}{2}\right]$ (please note that the arccosine of a positive number is a first quadrant angle)

- φ is the relative phase in $[0, 2\pi)$, neglecting the global phase $\varphi = \arg(\beta) - \arg(\alpha)$

This change of variables results in the following form of the state vector (or wave function) for a qubit on the Bloch sphere:

$$|\psi\rangle = \cos\frac{\theta}{2}|0\rangle + e^{i\varphi}\sin\frac{\theta}{2}|1\rangle$$

where we have replaced the spin-up and spin-down functions with the state vectors $|0\rangle$ and $|1\rangle$, respectively. From a chemical perspective, please note that the qubit state $|0\rangle$ indicates the angular momentum quantum numbers $l = 1, m_l = 0$ with spin-up, and therefore does not have any angular momentum projection on the z-axis. Furthermore, the qubit state $|1\rangle$ indicates the angular momentum quantum numbers $l = 1, m_l = 1$ with spin-down and does have angular momentum projection on the z-axis. This is important to remember when we introduce the Pauli matrices in *Section 3.1.6, Pauli matrices*, as we will see how the chemical information is modified when we apply operations.

On the Bloch sphere, angles are twice as big as in Hilbert space. For instance, $|0\rangle$ and $|1\rangle$ are orthogonal in Hilbert space, and on the Bloch sphere their angle is π. Further, we would like to point out that $\frac{\theta}{2}$ determines the probability to measure the $|0\rangle$ and $|1\rangle$ states with the following:

- $|0\rangle$ such that $P(0) = \cos^2 \frac{\theta}{2}$
- $|1\rangle$ such that $P(1) = \sin^2 \frac{\theta}{2}$

We show the Bloch vector for a qubit in different directions on the Bloch sphere, which we call pole states, as shown in the table shown in Figure 3.3:

Pole state	θ	φ	Bloch Vector	Direction
$|0\rangle$	0	Arbitrary	$\begin{pmatrix} 0 \\ 0 \\ 1 \end{pmatrix}$	"Up" on z-axis
$|1\rangle$	π	Arbitrary	$\begin{pmatrix} 0 \\ 0 \\ -1 \end{pmatrix}$	"Down" on z-axis
$|+\rangle = \frac{1}{\sqrt{2}}|0\rangle + \frac{1}{\sqrt{2}}|1\rangle$	$\frac{\pi}{2}$	0	$\begin{pmatrix} 1 \\ 0 \\ 0 \end{pmatrix}$	"Forward" on x-axis
$|-\rangle = \frac{1}{\sqrt{2}}|0\rangle - \frac{1}{\sqrt{2}}|1\rangle$	$\frac{\pi}{2}$	π	$\begin{pmatrix} -1 \\ 0 \\ 0 \end{pmatrix}$	"Backward" on x-axis
$|i+\rangle = \frac{1}{\sqrt{2}}|0\rangle + \frac{i}{\sqrt{2}}|1\rangle$	$\frac{\pi}{2}$	$\frac{\pi}{2}$	$\begin{pmatrix} 0 \\ 1 \\ 0 \end{pmatrix}$	"Right" on y-axis
$|i-\rangle = \frac{1}{\sqrt{2}}|0\rangle - \frac{i}{\sqrt{2}}|1\rangle$	$\frac{\pi}{2}$	$\frac{3\pi}{2}$	$\begin{pmatrix} 0 \\ -1 \\ 0 \end{pmatrix}$	"Left" on y-axis

Figure 3.3 – Pole states in the computational basis and their representation on the Bloch sphere

3.1.5. Displaying the Bloch vector corresponding to a state vector

In the following code, the `check` function performs sanity checks on a given complex vector (α, β) to ensure it is a state vector:

```
_EPS = 1e-
10 # Global variable used to chop small numbers to zero
```

```
def check(s):
  num_qubits = math.log2(len(s))
  # Check if param is a power of 2
  if num_qubits == 0 or not num_qubits.is_integer():
      raise Exception("Input complex vector length is not a
positive power of 2.")
  num_qubits = int(num_qubits)
  if num_qubits > 1:
      raise Exception("Only one complex vector is allowed as
input.")
  # Check if probabilities (amplitudes squared) sum to 1
  if not math.isclose(sum(np.absolute(s) ** 2), 1.0, abs_tol=_
EPS):
      raise Exception("Norm of complex vector does not equal
one.")
  return
```

Next, the `ToBloch()` function computes the Bloch vector of a given state vector (complex vector) and displays the angles in LaTeX format and the vector on the Bloch sphere. It has two input parameters:

- `s`: A state vector, a complex vector (α, β).

- `show`: Set to `True` to display the angles and the vector on the Bloch sphere.

It has three output parameters:

- `theta`: $\theta = 2\arccos(|\alpha|)$ in $[0, \pi]$ is the angle on the Bloch sphere.

- `phi`: $\varphi = \arg(\beta) - \arg(\alpha)$ in $[0, 2\pi]$ is the relative phase, neglecting the global phase.

- `r`: This is the vector on the Bloch sphere.

```
def ToBloch(s, show=True):
  check(s)
  phi = cmath.phase(s[1]) - cmath.phase(s[0])
  theta = 2*math.acos(abs(s[0]))
  r1 = math.sin(theta)*math.cos(phi)
  r2 = math.sin(theta)*math.sin(phi)
  r3 = math.cos(theta)
  r = (r1,r2,r3)
```

```
    if show:
        display(array_to_latex(s, prefix="\\
text{s} = ", precision = 2))
        display(array_to_latex([theta, phi], prefix="\\
text{theta, phi} = ", precision = 2))
        display(array_to_latex(r, prefix="\\
text{r} = ", precision = 2))
        b = qutip.Bloch()
        b.add_vectors(r)
        display(b.render())
    return theta, phi, r
```

The following code displays the Bloch vector corresponding to the state vector
$s = \left[\frac{1}{\sqrt{2}}, \frac{1}{2}(1+i)\right]$:

```
s = [1/math.sqrt(2),complex(0.5, 0.5)]
(theta, phi, r) = ToBloch(s)
```

Here is the result with the Bloch sphere displayed using the QuTiP Bloch() function:

$$s = \left[\frac{1}{\sqrt{2}}, \frac{1}{2}(1+i)\right]$$

$$\text{theta, phi} = [1,57, 0.79]$$

$$r = \left[\frac{1}{\sqrt{2}} \quad \frac{1}{\sqrt{2}} \quad 0\right]$$

Figure 3.4 – Displaying a vector on the Bloch sphere

Finally, the ToS function computes a state vector of a Bloch vector. It has three input parameters:

- theta: $\theta = 2\arccos(|\alpha|)$ in $[0, \pi]$ is the angle on the Bloch sphere.
- phi: $\varphi = \arg(\beta) - \arg(\alpha)$ in $[0, 2\pi]$ is the relative phase, neglecting the global phase.
- show: This is set to True to display the input angles and the state vector.

It has one output parameter:

- s: a state vector, a complex vector $\left(\cos\dfrac{\theta}{2}, e^{i\varphi}\sin\dfrac{\theta}{2}\right)$

```
def ToS(theta, phi, show=True):
    s = [math.cos(theta/2), complex(math.cos(phi) * math.
sin(theta/2), math.sin(phi) * math.sin(theta/2))]
    if show:
        display(array_to_latex([theta, phi], prefix="\\
text{theta, phi} = ", precision = 2))
        display(array_to_latex(s, prefix="\\
text{s} = ", precision = 1))
    return s
```

Here, we compute the complex amplitudes of a Bloch vector with $\theta = \pi/2$ and $\varphi = \pi/4$:

```
s = ToS(np.pi/2, np.pi/4)
```

Here is the result:

$$\text{theta, phi} = [1{,}57 \quad 0.79]$$

$$s = \left[\frac{1}{\sqrt{2}} \quad \frac{1}{2}(1+i)\right]$$

3.1.6. Pauli matrices

There are three Pauli matrices, σ_x, σ_y, and σ_z:

$$\sigma_x = \begin{pmatrix} 0 & 1 \\ 1 & 0 \end{pmatrix}, \sigma_y = \begin{pmatrix} 0 & -i \\ i & 0 \end{pmatrix}, \sigma_z = \begin{pmatrix} 1 & 0 \\ 0 & -1 \end{pmatrix}$$

which are Hermitian and unitary, making the square of each equal to the (2×2) identity matrix:

$$\sigma_x{}^2 = \sigma_y{}^2 = \sigma_z{}^2 = \begin{pmatrix} 1 & 0 \\ 0 & 1 \end{pmatrix}$$

Each of the Pauli matrices is equal to its inverse:

$$\sigma_x = \sigma_x^{-1}$$
$$\sigma_y = \sigma_y^{-1}$$
$$\sigma_z = \sigma_z^{-1}$$

We summarize the Pauli matrices and the operations on a qubit that yields the associated eigenvectors in *Figure 3.5*:

Pauli matrix	Eigenvector
$\sigma_x = \begin{pmatrix} 0 & 1 \\ 1 & 0 \end{pmatrix}$ Sign basis: $\{|+\rangle, |-\rangle\}$	$|+\rangle = \frac{1}{\sqrt{2}}|0\rangle + \frac{1}{\sqrt{2}}|1\rangle$
	$|-\rangle = \frac{1}{\sqrt{2}}|0\rangle - \frac{1}{\sqrt{2}}|1\rangle$
$\sigma_y = \begin{pmatrix} 0 & -i \\ i & 0 \end{pmatrix}$ Complex basis: $\{|i+\rangle, |i-\rangle\}$	$|i+\rangle = \frac{1}{\sqrt{2}}|0\rangle + \frac{i}{\sqrt{2}}|1\rangle$
	$|i-\rangle = \frac{1}{\sqrt{2}}|0\rangle - \frac{i}{\sqrt{2}}|1\rangle$
$\sigma_z = \begin{pmatrix} 1 & 0 \\ 0 & -1 \end{pmatrix}$ Standard basis: $\{|0\rangle, |1\rangle\}$	$|0\rangle$
	$|1\rangle$

Figure 3.5 – Pauli matrices and the associated eigenvectors

In *Figure 3.6*, we display on the far-left side the $|0\rangle$ qubit state, which has zero angular momentum projection on the z-axis, as indicated by the dark circle under the Bloch sphere. In the middle, we display the $|1\rangle$ qubit state, which has an angular momentum projection on the z-axis indicated by a light grey circle under the Bloch sphere. Recall we discussed the angular momentum projection for the $|0\rangle$ and $|1\rangle$ qubit states in *Section 3.1.4, Bloch sphere*. On the far-right side, we indicate the σ_z operation on the $|1\rangle$ qubit state, which modifies the angular momentum projection by π.

$$|\psi\rangle = \sqrt{1.00}\,|0\rangle + (\sqrt{0.00}\,)e^{i0}\,|1\rangle \quad |\psi\rangle = \sqrt{0.00}\,|0\rangle + (\sqrt{1.00}\,)e^{i0}\,|1\rangle \quad |\psi\rangle = \sqrt{0.00}\,|0\rangle + (\sqrt{1.00}\,)e^{i\pi}\,|1\rangle$$

Figure 3.6 – Pauli Z operation on basis states $|0\rangle$ and $|1\rangle$ illustrated with Grok the Bloch sphere

The σ_x operation does not have an effect on the angular momentum projection, while the σ_y operation modifies the angular momentum projection by $\pi/2$.

Measurement in the sign basis $\{|+\rangle, |-\rangle\}$

Let us measure a state $|\psi\rangle = \alpha|0\rangle + \beta|1\rangle$ in the sign basis $\{|+\rangle, |-\rangle\}$ which is also known as a measurement according to the Pauli matrix σ_x. To make this measurement, we perform a change of basis from the sign basis $\{|+\rangle, |-\rangle\}$ to the standard basis $\{|0\rangle, |1\rangle\}$ where:

$$|0\rangle = \frac{\sqrt{2}}{2}(|+\rangle + |-\rangle)$$

$$|1\rangle = \frac{\sqrt{2}}{2}(|+\rangle - |-\rangle)$$

which allows us to rewrite the state:

$$|\psi\rangle = \frac{\alpha + \beta}{\sqrt{2}}|+\rangle + \frac{\alpha - \beta}{\sqrt{2}}|-\rangle$$

The possible outcomes of a measurement with their corresponding probabilities and new state are listed in *Figure 3.7*:

Outcome of a measurement in the Sign basis $\{|+\rangle, |-\rangle\}$	Probability	New state
1	$\frac{1}{2}|\alpha + \beta|^2$	$|+\rangle$
-1	$\frac{1}{2}|\alpha - \beta|^2$	$|-\rangle$

Figure 3.7 – Measurement in the sign basis $\{|+\rangle, |-\rangle\}$

The expectation value of the measurement of a state $|\psi\rangle$ according to the Pauli σ_x operation is:

$$\langle\sigma_x\rangle_\psi = \langle\psi|\sigma_x|\psi\rangle$$

which means we need to calculate the bra in the sign basis by taking the complex conjugate transpose:

$$\langle\psi| = (|\psi\rangle^*)^T = \frac{\alpha^* + \beta^*}{\sqrt{2}}\langle+| + \frac{\alpha^* - \beta^*}{\sqrt{2}}\langle-|$$

We also need to transform the Pauli operation from the standard basis to the sign basis:

$$\sigma_x = 1\,|+\rangle\langle+| + (-1)\,|-\rangle\langle-| = \begin{pmatrix} 1 & 0 \\ 0 & -1 \end{pmatrix}$$

Recall that a ket times a bra, as seen previously ($|+\rangle\langle+|$ and $|-\rangle\langle-|$), is an outer product that yields a matrix, whereas a bra times a ket is a scalar. With this, we have the expectation value calculated as:

$$\langle\sigma_x\rangle_\psi = \langle\psi|\sigma_x|\psi\rangle = \frac{1}{2}(\alpha^* + \beta^*, \alpha^* - \beta^*)\begin{pmatrix} 1 & 0 \\ 0 & -1 \end{pmatrix}\begin{pmatrix} \alpha + \beta \\ \alpha - \beta \end{pmatrix}$$

$$= \frac{1}{2}(\alpha^* + \beta^*, (-1)(\alpha^* - \beta^*))\begin{pmatrix} \alpha + \beta \\ \alpha - \beta \end{pmatrix}$$

$$\langle\sigma_x\rangle_\psi = \frac{1}{2}(|\alpha + \beta|^2 - |\alpha - \beta|^2)$$

Please remember that in general, α and β are complex numbers, and the imaginary part can be zero. The expectation value is the sum of all the possible outcomes (1 and -1) of a measurement of a state $|\psi\rangle$ in the sign basis weighted by their probabilities.

Decomposing a matrix into the weighted sum of the tensor product of Pauli matrices

It can be shown that any matrix can be decomposed into the weighted sum of the tensor product of the identity matrix and the Pauli matrices $P_i = \otimes_j^N \sigma_{i,j}$, where $\sigma_{i,j} \in \{\mathbb{1}, \sigma_x, \sigma_y, \sigma_z\}$ with weights h_i and N qubits:

$$M = \sum_{i=1}^n h_i \otimes_j^N \sigma_{i,j}$$

For Hermitian matrices, all weights h_i are real.

We provide a proof for any 2x2 matrix, $M = \begin{pmatrix} m_{0,0} & m_{0,1} \\ m_{1,0} & m_{1,1} \end{pmatrix}$:

$$\sigma_z \sigma_x = \begin{pmatrix} 1 & 0 \\ 0 & -1 \end{pmatrix}\begin{pmatrix} 0 & 1 \\ 1 & 0 \end{pmatrix} = \begin{pmatrix} 0 & 1 \\ -1 & 0 \end{pmatrix} = i\sigma_y$$

$$\sigma_x \sigma_z = \begin{pmatrix} 0 & 1 \\ 1 & 0 \end{pmatrix}\begin{pmatrix} 1 & 0 \\ 0 & -1 \end{pmatrix} = \begin{pmatrix} 0 & -1 \\ 1 & 0 \end{pmatrix} = -i\sigma_y$$

$$\frac{\mathbb{1} + \sigma_z}{2} = \frac{1}{2}\left(\begin{pmatrix} 1 & 0 \\ 0 & 1 \end{pmatrix} + \begin{pmatrix} 1 & 0 \\ 0 & -1 \end{pmatrix}\right) = \begin{pmatrix} 1 & 0 \\ 0 & 0 \end{pmatrix} = |0\rangle\langle0|$$

$$\frac{\mathbb{1} - \sigma_z}{2} = \frac{1}{2}\left(\begin{pmatrix} 1 & 0 \\ 0 & 1 \end{pmatrix} - \begin{pmatrix} 1 & 0 \\ 0 & -1 \end{pmatrix}\right) = \begin{pmatrix} 0 & 0 \\ 0 & 1 \end{pmatrix} = |1\rangle\langle1|$$

Since $\sigma_x|0\rangle = |1\rangle$, hence $\langle 1| = \langle 0|\sigma_x$, we have:

$$|0\rangle\langle 1| = |0\rangle\langle 0|\sigma_x = \frac{\mathbb{1} + \sigma_z}{2}\sigma_x = \frac{\sigma_x + i\sigma_y}{2}$$

$$|1\rangle\langle 0| = \sigma_x|0\rangle\langle 0| = \sigma_x\frac{\mathbb{1} + \sigma_z}{2} = \frac{\sigma_x - i\sigma_y}{2}$$

Starting from the decomposition of a 2x2 matrix as a sum of outer products:

$$M = \begin{pmatrix} m_{0,0} & m_{0,1} \\ m_{1,0} & m_{1,1} \end{pmatrix} = m_{0,0}|0\rangle\langle 0| + m_{0,1}|0\rangle\langle 1| + m_{1,0}|1\rangle\langle 0| + m_{1,1}|1\rangle\langle 1|$$

we can then write:

$$M = m_{0,0}\frac{\mathbb{1} + \sigma_z}{2} + m_{0,1}\frac{\sigma_x + i\sigma_y}{2} + m_{1,0}\frac{\sigma_x - i\sigma_y}{2} + m_{1,1}\frac{\mathbb{1} - \sigma_z}{2}$$

$$M = \frac{m_{0,0} + m_{1,1}}{2}\mathbb{1} + \frac{m_{0,1} + m_{1,0}}{2}\sigma_x + i\frac{m_{0,1} - m_{1,0}}{2}\sigma_y + \frac{m_{0,0} - m_{1,1}}{2}\sigma_z$$

3.2. Quantum gates

Quantum gates are unitary operators ($U^\dagger U = UU^\dagger = \mathbb{1}$) working on one, two, or three qubits. The norm is preserved when applied to a quantum state. The action of a quantum gate on a quantum state corresponds to the multiplication of the matrix representing the gate by the vector representing the quantum state: $U|q\rangle$.

In this section, a tensor product of n qubits is represented with the first qubit on the left-most side of the tensor product: $|q\rangle = |q_0\rangle|q_1\rangle \ldots |q_{n-1}\rangle = |q_0, q_1, \ldots, q_{n-1}\rangle = \bigotimes_{i=0}^{n-1} |q_i\rangle$ where $q_i \in \{0,1\}$. Please note that we are not using the Qiskit tensor ordering of qubits unless specifically specified.

In this section, we cover the following:

- *Section 3.2.1, Single-qubit quantum gates*
- *Section 3.2.2, Two-qubit quantum gates*
- *Section 3.2.3, Three-qubit quantum gates*
- *Section 3.2.4, Serial wired gates and parallel quantum gates*
- *Section 3.2.5, Creation of a Bell state*
- *Section 3.2.6, Parallel Hadamard gates*

3.2.1. Single-qubit quantum gates

A single-qubit gate U has a (2×2) unitary matrix form: $U^\dagger U = UU^\dagger = \mathbb{1}$.

In this section, we describe in detail the following:

- X gate
- Hadamard (H) gate
- Generalized single-qubit quantum gate

We summarize commonly used quantum gates as well as provide some useful relationships.

X gate

An X gate maps $|0\rangle$ to $|1\rangle$ and $|1\rangle$ to $|0\rangle$. It is the quantum equivalent of the NOT gate for classical computers and is sometimes called a bit-flip. For classical computing, the NOT gate changes a 0 to a 1 and a 1 to a 0. The X gate equates to a rotation by π radians around the X-axis of the Bloch sphere.

$$X = \begin{pmatrix} 0 & 1 \\ 1 & 0 \end{pmatrix}$$

Hadamard (H) gate

A Hadamard gate maps the basis state $|0\rangle$ to $\frac{1}{\sqrt{2}}(|0\rangle + |1\rangle)$, which is also written as $|+\rangle$, and $|1\rangle$ to $\frac{1}{\sqrt{2}}(|0\rangle - |1\rangle)$, which is also written as $|-\rangle$. It represents a rotation of π about the axis that is in the middle (45° angle) of the x- and z-axis. A measurement of the state $|+\rangle$ or of the state $|-\rangle$ will have equal probabilities of being 0 or 1, creating a superposition of states.

$$H = \frac{1}{\sqrt{2}} \begin{pmatrix} 1 & 1 \\ 1 & -1 \end{pmatrix}$$

It is convenient to write the Hadamard gate applied to the 0th qubit (x_0) in the register as follows: $H|x_0\rangle = \frac{1}{\sqrt{2}}(|0\rangle + (-1)^{x_0}|1\rangle)$ where $x_0 \in \{0, 1\}$. Please note that the Hadamard gate (H) has similar notation to the Hamiltonian operator ($\hat{H}$); the difference is the hat ($\hat{}$) symbol.

General single-qubit quantum gate

All single-qubit gates can be obtained from the following matrix $u(\theta, \varphi, \lambda)$, which describes all unitary matrices up to a global phase factor by an appropriate choice of parameters θ, φ, λ with $\theta \in [0, \pi]$, $\varphi \in [0, 2\pi]$, $\lambda \in [0, 2\pi]$ [Qiskit_Op]:

$$u(\theta, \varphi, \lambda) = \begin{pmatrix} \cos\dfrac{\theta}{2} & -e^{i\lambda}\sin\dfrac{\theta}{2} \\ e^{i\varphi}\sin\dfrac{\theta}{2} & e^{i(\varphi+\lambda)}\cos\dfrac{\theta}{2} \end{pmatrix}$$

The gate $u\left(\dfrac{\pi}{2}, \varphi, \lambda\right)$ has the matrix form:

$$u\left(\frac{\pi}{2}, \varphi, \lambda\right) = \frac{1}{\sqrt{2}}\begin{pmatrix} 1 & -e^{i\lambda} \\ e^{i\varphi} & e^{i(\varphi+\lambda)} \end{pmatrix}$$

The gate $p(\lambda) = u(0,0,\lambda)$ and has the matrix form:

$$p(\lambda) = \begin{pmatrix} 1 & 0 \\ 0 & e^{i\lambda} \end{pmatrix}$$

Summary of single-qubit quantum gates and useful relationships

Figure 3.8 presents the main list of single-qubit quantum gates:

Operator	Gate	(2×2) unitary matrix	Maps $\lvert 0\rangle$ to	Maps $\lvert 1\rangle$ to	Maps $\lvert +\rangle$ to	Maps $\lvert -\rangle$ to
Identity (I, 𝟙)		$\begin{pmatrix} 1 & 0 \\ 0 & 1 \end{pmatrix}$	$\lvert 0\rangle$	$\lvert 1\rangle$	$\lvert +\rangle$	$\lvert -\rangle$
Pauli-X (σ_X, X)	X	$\begin{pmatrix} 0 & 1 \\ 1 & 0 \end{pmatrix}$	$\lvert 1\rangle$	$\lvert 0\rangle$	$\lvert +\rangle$	$-\lvert -\rangle$
Pauli-Y (σ_Y, Y)	Y	$\begin{pmatrix} 0 & -i \\ i & 0 \end{pmatrix}$	$i\lvert 1\rangle$	$-i\lvert 0\rangle$		
Pauli-Z (σ_Z, Z)	Z	$\begin{pmatrix} 1 & 0 \\ 0 & -1 \end{pmatrix}$	$\lvert 0\rangle$	$-\lvert 1\rangle$	$\lvert -\rangle$	$\lvert +\rangle$
Hadamard (H)	H	$\dfrac{1}{\sqrt{2}}\begin{pmatrix} 1 & 1 \\ 1 & -1 \end{pmatrix}$	$\lvert +\rangle$	$\lvert -\rangle$	$\lvert 0\rangle$	$\lvert 1\rangle$
$S \sqrt{Z} \, p\left(\dfrac{\pi}{2}\right)$	S	$\begin{pmatrix} 1 & 0 \\ 0 & i \end{pmatrix}$	$\lvert 0\rangle$	$i\lvert 1\rangle$		
$S^\dagger \sqrt{Z}^\dagger \, p\left(-\dfrac{\pi}{2}\right)$	$S^\dagger$	$\begin{pmatrix} 1 & 0 \\ 0 & -i \end{pmatrix}$	$\lvert 0\rangle$	$-i\lvert 1\rangle$		
$T \sqrt{S} \, p\left(\dfrac{\pi}{4}\right)$	T	$\begin{pmatrix} 1 & 0 \\ 0 & e^{i\frac{\pi}{4}} \end{pmatrix}$	$\lvert 0\rangle$	$e^{i\frac{\pi}{4}}\lvert 1\rangle$		
$T^\dagger \sqrt{S}^\dagger \, p\left(-\dfrac{\pi}{4}\right)$	$T^\dagger$	$\begin{pmatrix} 1 & 0 \\ 0 & e^{-i\frac{\pi}{4}} \end{pmatrix}$	$\lvert 0\rangle$	$e^{-i\frac{\pi}{4}}\lvert 1\rangle$		

Figure 3.8 – Single-qubit quantum gates

The rotation operator gates RX, RY, and RZ perform rotations about the X, Y, and Z axes respectively of the Bloch sphere:

$$RX(\theta) = \exp\left(-i\frac{\theta}{2}X\right) = \begin{pmatrix} \cos\frac{\theta}{2} & -i\sin\frac{\theta}{2} \\ -i\sin\frac{\theta}{2} & \cos\frac{\theta}{2} \end{pmatrix}$$

$$RY(\theta) = \exp\left(-i\frac{\theta}{2}Y\right) = \begin{pmatrix} \cos\frac{\theta}{2} & -\sin\frac{\theta}{2} \\ \sin\frac{\theta}{2} & \cos\frac{\theta}{2} \end{pmatrix}$$

$$RZ(\lambda) = \exp\left(-i\frac{\lambda}{2}Z\right) = \begin{pmatrix} -e^{i\frac{\lambda}{2}} & 0 \\ 0 & e^{i\frac{\lambda}{2}} \end{pmatrix}$$

We would like to point out that the X gate can be obtained by using a combination of the Hadamard gate and the Z gate: $X = HZH$. The converse is also true: $Z = HXH$. It means we can project a state onto the X-axis of the Bloch sphere when applying a H gate before measuring. Same with the Y-axis when applying first an $S^\dagger$ gate, then an H gate. This way we can perform qubit tomography (that is, reconstructing the Bloch vector through X, Y, and Z measurements). These gate operations are summarized in *Figure 3.9*:

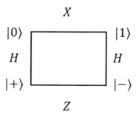

Figure 3.9 – Relations $X = HZH$, $Z = HXH$

3.2.2. Two-qubit quantum gates

A two-qubit gate U is a (4×4) unitary matrix, $U^\dagger U = UU^\dagger = \mathbb{1}$ that acts on two qubits.

We summarize commonly used two-qubit quantum gates in *Figure 3.10*.

Operator	Gate	4x4 unitary matrix	Description
Controlled Not (CNOT, CX)	CX	$\begin{pmatrix} 1 & 0 & 0 & 0 \\ 0 & 1 & 0 & 0 \\ 0 & 0 & 0 & 1 \\ 0 & 0 & 1 & 0 \end{pmatrix}$	If the first qubit is $\lvert 1 \rangle$ it performs the Pauli-X (NOT) operation on the second qubit, otherwise it leaves it unchanged.
Controlled Z (CZ)	CZ	$\begin{pmatrix} 1 & 0 & 0 & 0 \\ 0 & 1 & 0 & 0 \\ 0 & 0 & 1 & 0 \\ 0 & 0 & 0 & -1 \end{pmatrix}$	If the first qubit is $\lvert 1 \rangle$ then it performs the Z operation on the second qubit, otherwise it leaves it unchanged.
Controlled-U (CU)	CU	$\begin{pmatrix} 1 & 0 & 0 & 0 \\ 0 & 1 & 0 & 0 \\ 0 & 0 & u_{00} & u_{01} \\ 0 & 0 & u_{10} & u_{11} \end{pmatrix}$	If the first qubit is $\lvert 1 \rangle$ then it performs the $U = \begin{pmatrix} u_{00} & u_{01} \\ u_{10} & u_{11} \end{pmatrix}$ operation on the second qubit, otherwise it leaves it unchanged.
SWAP	SWAP	$\begin{pmatrix} 1 & 0 & 0 & 0 \\ 0 & 0 & 1 & 0 \\ 0 & 1 & 0 & 0 \\ 0 & 0 & 0 & 1 \end{pmatrix}$	It swaps two qubits.

Figure 3.10 – Two-qubit quantum gates

3.2.3. Three-qubit quantum gates

A three-qubit gate U is an (8×8) unitary matrix $U^\dagger U = UU^\dagger = \mathbb{1}$ that acts on three qubits. We summarize commonly used three-qubit quantum gates in *Figure 3.11*:

Operator	Gate	8x8 unitary matrix	Description
Toffoli gate (CCX)	CCX	$\begin{pmatrix} 1 & 0 & 0 & 0 & 0 & 0 & 0 & 0 \\ 0 & 1 & 0 & 0 & 0 & 0 & 0 & 0 \\ 0 & 0 & 1 & 0 & 0 & 0 & 0 & 0 \\ 0 & 0 & 0 & 1 & 0 & 0 & 0 & 0 \\ 0 & 0 & 0 & 0 & 1 & 0 & 0 & 0 \\ 0 & 0 & 0 & 0 & 0 & 1 & 0 & 0 \\ 0 & 0 & 0 & 0 & 0 & 0 & 0 & 1 \\ 0 & 0 & 0 & 0 & 0 & 0 & 1 & 0 \end{pmatrix}$	If the first two qubits are in state $\lvert 1 \rangle$ then it performs the Pauli-X (NOT) operation on the third qubit, otherwise it leaves it unchanged.
Controlled Swap	CSWAP	$\begin{pmatrix} 1 & 0 & 0 & 0 & 0 & 0 & 0 & 0 \\ 0 & 1 & 0 & 0 & 0 & 0 & 0 & 0 \\ 0 & 0 & 1 & 0 & 0 & 0 & 0 & 0 \\ 0 & 0 & 0 & 0 & 0 & 1 & 0 & 0 \\ 0 & 0 & 0 & 0 & 1 & 0 & 0 & 0 \\ 0 & 0 & 0 & 1 & 0 & 0 & 0 & 0 \\ 0 & 0 & 0 & 0 & 0 & 0 & 1 & 0 \\ 0 & 0 & 0 & 0 & 0 & 0 & 0 & 1 \end{pmatrix}$	If the first qubit is $\lvert 1 \rangle$ then the controlled swap gate exchanges the second and third qubits.

Figure 3.11 – Three-qubit quantum gates

3.2.4. Serially wired gates and parallel quantum gates

Operations on quantum gates are applied sequentially from left to right, and there are no loops. Two gates U and V in series are equivalent to the matrix product of the two gates, as shown in *Figure 3.12*:

Two gates in series	Input	Equivalent matrix product	Output
$\lvert x \rangle - \mathrm{U} - \mathrm{V} -$	$\lvert x \rangle$	$- VU -$	$VU \lvert x \rangle$

Figure 3.12 – Serially wired quantum gates

Two gates U and V in parallel are equivalent to the tensor product of the two gates $U \otimes V$, as shown in *Figure 3.13*:

Two gates in parallel	Input	Equivalent tensor product	Output
$\lvert x \rangle - \mathrm{U} -$ $\lvert y \rangle - \mathrm{V} -$	$\lvert x \rangle$ $\lvert y \rangle$	$- U \otimes V -$	$U\lvert x \rangle \otimes U\lvert y \rangle = (U \otimes V)\lvert x \rangle \otimes \lvert y \rangle$
$\lvert x \rangle - - -$ $\lvert y \rangle - \mathrm{V} -$	$\lvert x \rangle$ $\lvert y \rangle$	$- \mathbb{1} \otimes V -$	$(\mathbb{1} \otimes V)\lvert x \rangle \otimes \lvert y \rangle$

Figure 3.13 – Parallel quantum gates

3.2.5. Creation of a Bell state

Bell states are maximally entangled pure quantum states, and there are only four:

$$\lvert \Phi_+ \rangle = \frac{1}{\sqrt{2}}(\lvert 00 \rangle + \lvert 11 \rangle)$$

$$\lvert \Phi_- \rangle = \frac{1}{\sqrt{2}}(\lvert 00 \rangle - \lvert 11 \rangle)$$

$$\lvert \Psi_+ \rangle = \frac{1}{\sqrt{2}}(\lvert 01 \rangle + \lvert 10 \rangle)$$

$$\lvert \Psi_- \rangle = \frac{1}{\sqrt{2}}(\lvert 01 \rangle - \lvert 10 \rangle)$$

A quantum circuit is an ordered sequence of instructions, quantum gates, measurements, and resets that is applied to registers of qubits and may be conditioned on real-time classical computation. Several quantum hardware platforms now support dynamic quantum circuits, which allow concurrent classical processing of mid-circuit measurement results [Corcoles] [IBM_mid]. In *Section 3.4, Preparing a permutation symmetric or antisymmetric state*, we demonstrate a classical program that aims to obtain the desired quantum state by post selecting the result of a measurement of a control qubit. There is no loop in a quantum circuit, but we can have a classical loop that appends a quantum sub-circuit. In Qiskit, we use the `QuantumRegister` class to create a register of qubits and the `QuantumCircuit` class to create a quantum circuit.

Let's build a quantum circuit that creates the first Bell state $|\Phi_+\rangle$ with Qiskit:

```
q = QuantumRegister(2)
qc = QuantumCircuit(q)


qc.h(q[0])
qc.cx(q[0], q[1])


qc.draw(output='mpl')
```

Figure 3.14 shows the result:

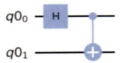

Figure 3.14 – Qiskit quantum circuit that creates a Bell state

We use the `Statevector.from_instruction()` class method from the `quantum_info` module to get the final state vector $|s\rangle$:

```
s = qi.Statevector.from_instruction(qc)
s.draw('latex', prefix='|s \\rangle = ')
```

Figure 3.15 shows the result:

$$|s\rangle = \begin{bmatrix} \frac{1}{\sqrt{2}} & 0 & 0 & \frac{1}{\sqrt{2}} \end{bmatrix}$$

Figure 3.15 – Final state vector – Bell state

The final state vector can only be measured in either the state $|00\rangle$ or $|11\rangle$, each with a probability of 1/2.

We use the `DensityMatrix.from_instruction()` class method to obtain the density matrix representation of the final state vector:

```
rho = qi.DensityMatrix.from_instruction(qc)
rho.draw('latex', prefix='\\rho = ')
```

Figure 3.16 shows the result:

$$\rho = \begin{bmatrix} \frac{1}{2} & 0 & 0 & \frac{1}{2} \\ 0 & 0 & 0 & 0 \\ 0 & 0 & 0 & 0 \\ \frac{1}{2} & 0 & 0 & \frac{1}{2} \end{bmatrix}$$

Figure 3.16 – Density matrix – Bell state

We can visualize the density matrix using a cityscape plot of the state:

```
from qiskit.visualization import plot_state_city
plot_state_city(rho.data, title='Density Matrix')
```

Figure 3.17 shows the result:

Density Matrix

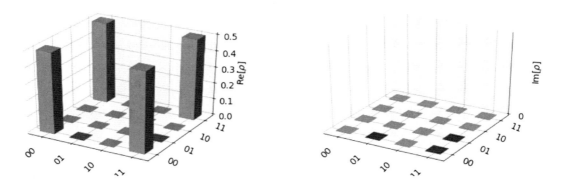

Figure 3.17 – Cityscape plot – Bell state

3.2.6. Parallel Hadamard gates

It can be shown that applying parallel Hadamard gates to a register of qubits initialized in the zero state puts it in a uniform superposition of all possible states. Let's experiment with the effect of applying one, two, and three Hadamard gates. In this section, we'll build the following:

- One Hadamard gate

- Two parallel Hadamard gates

- Three parallel Hadamard gates

The create_parallel_H() function creates a quantum circuit with n parallel Hadamard gates:

```
def create_parallel_H(n):
    q = QuantumRegister(n, 'q')
    qc = QuantumCircuit(q)
    for k in range(n):
        qc.h(k)
    return qc
```

The function run_parallel_H() creates and executes a quantum circuit with n parallel Hadamard gates and displays a diagram of the final state vector:

```
def run_parallel_H(n):
  qc = create_parallel_H(n)
  s = qi.Statevector.from_instruction(qc)
  display(s.draw('latex'))
  display(qc.draw(output='mpl'))
  return
```

Let's create a quantum circuit with just one Hadamard gate:

```
run_parallel_H(1)
```

Figure 3.18 shows the result:

$$\frac{\sqrt{2}}{2}|0\rangle + \frac{\sqrt{2}}{2}|1\rangle$$

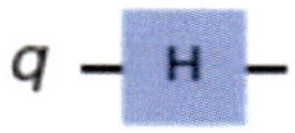

$$q - \boxed{H} -$$

Figure 3.18 – One Hadamard gate

Next, we build a quantum circuit with two parallel Hadamard gates:

```
run_parallel_H(2)
```

Figure 3.19 shows the result:

$$\frac{1}{2}|00\rangle + \frac{1}{2}|01\rangle + \frac{1}{2}|10\rangle + \frac{1}{2}|11\rangle$$

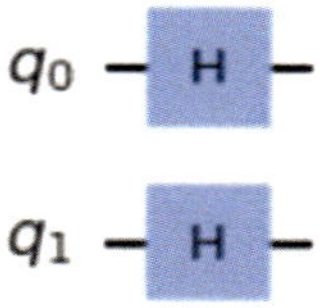

$$q_0 - \boxed{H} -$$

$$q_1 - \boxed{H} -$$

Figure 3.19 – Two parallel Hadamard gates

Last, let's build a circuit with three parallel Hadamard gates:

```
run_parallel_H(3)
```

Figure 3.20 shows the result:

$$\frac{\sqrt{2}}{4}|000\rangle + \frac{\sqrt{2}}{4}|001\rangle + \frac{\sqrt{2}}{4}|010\rangle + \frac{\sqrt{2}}{4}|011\rangle + \frac{\sqrt{2}}{4}|100\rangle + \frac{\sqrt{2}}{4}|101\rangle + \frac{\sqrt{2}}{4}|110\rangle + \frac{\sqrt{2}}{4}|111\rangle$$

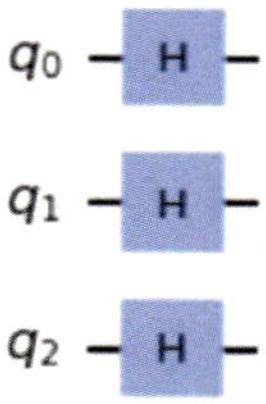

$$q_0 - \boxed{H} -$$

$$q_1 - \boxed{H} -$$

$$q_2 - \boxed{H} -$$

Figure 3.20 – Three parallel Hadamard gates

3.3. Computation-driven interference

In this section, we introduce the process of a generic quantum computation in *Section 3.3.1, Quantum computation process*. Then we give an example of a simulation inspired by a chemical experiment in *Section 3.3.2, Simulating interferometric sensing of a quantum superposition of left- and right-handed enantiomer states*. In chemistry, molecules or ions that are mirror images of each other are called enantiomers or optical isomers. If these images are non-superimposable, they are called chiral molecules [ChemChiral] and they differ in their ability to rotate plane polarized light either to the left or to the right [Wonders]. Researchers have proposed an experiment to prepare a quantum superposition of left- and right-handed states of enantiomers and to perform interferometric sensing of chirality-dependent forces [Stickler].

3.3.1. Quantum computation process

Quantum computing uses interference and the quantum physical phenomena of superposition and entanglement. A typical quantum computation comprises the following steps:

1. Prepare a uniform superposition of all possible basis states. A register of qubits initialized in the zero state is put in a uniform superposition of all possible basis states simply by applying parallel Hadamard gates, as we illustrated previously.

2. Orchestrate quantum interference and entanglement. A quantum algorithm ought to be designed such that at the end of a computation, only the relative amplitudes and the phases of those quantum states that are of interest will remain.

3. Repeat the measurements multiple times. Measurements are repeated hundreds or thousands of times in order to obtain a distribution over the possible measurement outcomes. This is the key difference between quantum and classical computing.

3.3.2. Simulating interferometric sensing of a quantum superposition of left- and right-handed enantiomer states

Let's design, with Qiskit, a quantum circuit inspired by the interferometer involving enantiomers. We represent a single enantiomer with two qubits. We encode in the direction of propagation qubit $|q_1\rangle$ the horizontal propagation as the state $|0\rangle$ and the vertical propagation as the state $|1\rangle$. We simulate a mirror by the Pauli σ_x matrix and a beam splitter (BS) by the matrix $\frac{1}{\sqrt{2}}(\mathbb{1} + i\sigma_x)$:

$$BS = \frac{1}{\sqrt{2}}(\mathbb{1} + i\sigma_x) = \frac{1}{\sqrt{2}}\left(\begin{pmatrix} 1 & 0 \\ 0 & 1 \end{pmatrix} + i\begin{pmatrix} 0 & 1 \\ 1 & 0 \end{pmatrix}\right) = \frac{1}{\sqrt{2}}\begin{pmatrix} 1 & i \\ i & 1 \end{pmatrix}$$

By convention, a phase shift of $\frac{\pi}{2}$ is assigned to reflection. From the preceding unitary matrix, we create a beam splitter gate named BS with the following Qiskit code:

```
from qiskit.extensions import UnitaryGate
i = complex(0.0, 1.0)
BS = 1/np.sqrt(2) * np.array([[1,i],[i,1]])
BS = UnitaryGate(BS, 'Beam Splitter')
```

We encode the following in the handedness qubit $|q_0\rangle$:

- A left-handed state as the $|0\rangle$ state

- A right-handed state as the $|1\rangle$ state

- A superposition of left- and right-handed states as $\frac{1}{\sqrt{2}}(|0\rangle + |1\rangle)$ obtained by applying a Hadamard gate

We simulate a polarizing beam splitter (PBS):

$$PBS = \begin{pmatrix} 1 & 0 & 0 & 0 \\ 0 & 0 & 0 & 1 \\ 0 & 0 & 1 & 0 \\ 0 & 1 & 0 & 0 \end{pmatrix}$$

which transmits left-handed and reflects right-handed states with the matrix PBS [Rioux]. Unlike the beam splitter, there is no phase change on reflection. From the preceding unitary matrix, we create a polarizing beam splitter gate named PBS with the following Qiskit code:

```
PBS = np.array([[1,0,0,0],[0,0,0,1],[0,0,1,0],[0,1,0,0]])
PBS = UnitaryGate(PBS, 'PBS')
```

We define the `show()` function, which displays the drawing of a quantum circuit and the state of the state vector using LaTeX as follows:

```
def show(qc):
    display(qc.draw(output='mpl'))
    s = qi.Statevector.from_instruction(qc)
    display(array_to_latex(s, prefix="\\
text{state vector} = ", precision = 2))
    return
```

We simulate the action of a polarizing beam splitter on an enantiomer moving horizontally and in the right-handed state with the following Qiskit code:

```
q = QuantumRegister(2, 'q') # register of 2 qubits
# q[0] handedness qubit, |0⟩ left-handed, |1⟩ right-handed
# q[1] direction of propagation qubit, |0⟩ horizontal, |1⟩
vertical
qc = QuantumCircuit(q)


qc.x([0]) # Right-handed
show(qc)


qc.append(PBS, q)
show(qc)
```

Figure 3.21 shows the result:

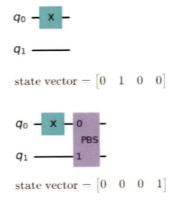

Figure 3.21 – Simulation of the interaction of a right-handed enantiomer with a polarizing beam splitter

The enantiomer moving horizontally and in the right-handed state, represented by $|q_1 q_0\rangle = |01\rangle$, using Qiskit tensor ordering of qubits, has been reflected in the vertical direction of propagation by the polarizing beam splitter, represented by $|q_1 q_0\rangle = |11\rangle$.

We simulate the action of a polarizing beam splitter on an enantiomer moving horizontally and in a superposition of left- and right-handed states with the following Qiskit code:

```
q = QuantumRegister(2, 'q') # register of 2 qubits
# q[0] handedness qubit,|0) left-handed, |1) right-handed
# q[1] direction of propagation qubit, |0) horizontal, |1)
vertical
qc = QuantumCircuit(q)


qc.h(q[0]) # Put enantiomer in a superposition of left- and
right-handed states
show(qc)


qc.append(PBS, q)
show(qc)
```

Figure 3.22 shows the result:

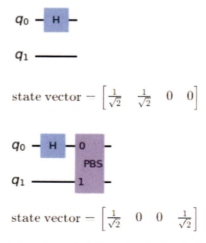

Figure 3.22 – Simulation of the interaction of a left- and right-handed enantiomer with a polarizing beam splitter

The enantiomer moving horizontally and in a superposition of the left- and right-handed states, represented by $|q_1 q_0\rangle = |0\rangle \otimes \frac{1}{\sqrt{2}}(|0\rangle + |1\rangle) = \frac{1}{\sqrt{2}}(|00\rangle + |01\rangle)$, using Qiskit tensor ordering of qubits, has been put by the polarizing beam splitter in the Bell state $|q_1 q_0\rangle = |\Phi_+\rangle = \frac{1}{\sqrt{2}}(|00\rangle + |11\rangle)$, a superposition of left-handed moving horizontally and right-handed moving vertically, thereby achieving interferometric sensing.

Now we move on to preparing permutation symmetric and antisymmetic states.

3.4. Preparing a permutation symmetric or antisymmetric state

Given two qubits $|q_1\rangle$ and $|q_2\rangle$, we want to build a symmetrized state that remains invariant under a permutation of the qubits $|q_1\rangle$ and $|q_2\rangle$, or an antisymmetrized state that is multiplied by -1 under a permutation of the qubits $|q_1\rangle$ and $|q_2\rangle$. In this section, we show how to prepare such states in a probabilistic manner with a quantum circuit prepared and simulated with Qiskit:

- *Section 3.4.1, Creating random states*
- *Section 3.4.2, Creating a quantum circuit and initializing qubits*
- *Section 3.4.3, Creating a circuit that swaps two qubits with a controlled swap gate*
- *Section 3.4.4, Post selecting the control qubit until the desired state is obtained*
- *Section 3.4.5, Examples of final symmetrized and antisymmetrized states*

3.4.1. Creating random states

We define a function called `init_random()` that creates random 1-qubit states `s1` and `s2` that we will use later to run experiments with random states:

```
def init_random():
  # Create random 1-qubit state s1
  s1 = qi.random_statevector(2)
  display(array_to_latex(s1, prefix="\\
text{State 1} =", precision=2))


  # Create random 1-qubit state s2
  s2 = qi.random_statevector(2)
```

```
  display(array_to_latex(s2, prefix="\\
text{State 2} =", precision =2))


  return s1, s2
```

3.4.2. Creating a quantum circuit and initializing qubits

We define the `setup_qc()` function, which sets up the initialization instruction to create qubits $|q_1\rangle$ and $|q_2\rangle$ from the state vectors `s1` and `s2` given as input and a quantum circuit `qc` with a control qubit $|q_0\rangle$ initialized in state $|0\rangle$, qubits $|q_1\rangle$ and $|q_2\rangle$, and a classical register `c` for measuring $|q_0\rangle$:

```
def setup_qc(s1, s2, draw=False):
  init_q1 = Initialize(s1)
  init_q1.label = "init_q1"


  init_q2 = Initialize(s2)
  init_q2.label = "init_q2"


  q = QuantumRegister(3, 'q') # register of 3 qubits
  c = ClassicalRegister(1, name="c") # and 1 classical register
  qc = QuantumCircuit(q,c)
  qc.append(init_q1, [1])
  qc.append(init_q2, [2])
  qc.barrier()
  if draw:
    display(qc.draw(output='mpl'))
  return qc, q, c
```

3.4.3. Creating a circuit that swaps two qubits with a controlled swap gate

We define the `swapper()` function, which creates a quantum circuit as follows [Spheres]:

- Applying a Hadamard gate to the control qubit $|q_0\rangle$ which puts it in the state $\frac{1}{\sqrt{2}}(|0\rangle + |1\rangle)$

- Applying a controlled swap gate, which puts the two qubits $|q_1\rangle$ and $|q_2\rangle$ in a superposition of being swapped and not swapped
- Applying again a Hadamard gate to the control qubit $|q_0\rangle$

Here is the implementation:

```
def swapper(draw=False):
  q = QuantumRegister(3, 'q') # register of 3 qubits
  qc = QuantumCircuit(q, name='Swapper')
  qc.h(q[0])
  qc.cswap(q[0], q[1], q[2])
  qc.h(q[0])
  if draw:
    print("Swapper circuit")
    display(qc.draw(output='mpl'))
  return qc
```

Let's get the unitary matrix corresponding to the swapper quantum circuit with the unitary simulator:

```
q = QuantumRegister(3, 'q') # register of 3 qubits
qc = QuantumCircuit(q)
qc.append(swapper(draw=True), qargs=q)


# Selecting the unitary_simulator
backend = Aer.get_backend('unitary_simulator')


# Executing the job and getting the result as an object
job = execute(qc, backend)
result = job.result()


# Getting the unitary matrix from the result object
U = result.get_unitary(qc, decimals=2)
array_to_latex(U, prefix="\\
text{swapper unitary} = ", precision = 2)
```

Figure 3.23 shows the result:

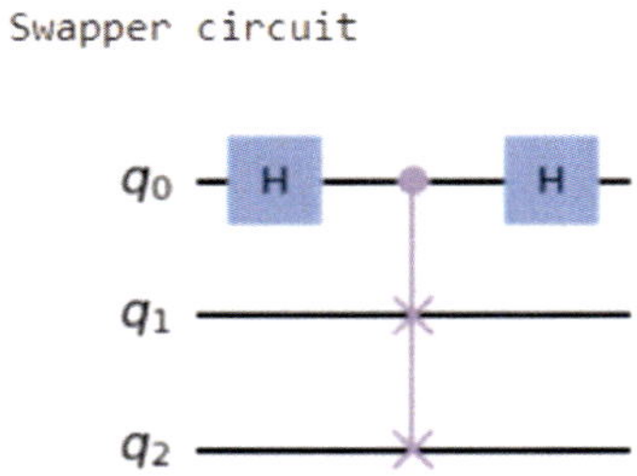

Swapper circuit

$$\text{swapper unitary} = \begin{bmatrix} 1 & 0 & 0 & 0 & 0 & 0 & 0 & 0 \\ 0 & 1 & 0 & 0 & 0 & 0 & 0 & 0 \\ 0 & 0 & \frac{1}{2} & \frac{1}{2} & \frac{1}{2} & -\frac{1}{2} & 0 & 0 \\ 0 & 0 & \frac{1}{2} & \frac{1}{2} & -\frac{1}{2} & \frac{1}{2} & 0 & 0 \\ 0 & 0 & \frac{1}{2} & -\frac{1}{2} & \frac{1}{2} & \frac{1}{2} & 0 & 0 \\ 0 & 0 & -\frac{1}{2} & \frac{1}{2} & \frac{1}{2} & \frac{1}{2} & 0 & 0 \\ 0 & 0 & 0 & 0 & 0 & 0 & 1 & 0 \\ 0 & 0 & 0 & 0 & 0 & 0 & 0 & 1 \end{bmatrix}$$

Figure 3.23 – Unitary matrix of the swapper circuit

Computing the action of the swapper unitary

The initial state vector pertaining to the two qubits $|q_1\rangle$ and $|q_2\rangle$ can be written as follows:

$$|q_2 q_1\rangle = \begin{pmatrix} x_0 \\ x_1 \\ x_2 \\ x_3 \end{pmatrix}$$

The swapper unitary acts on the initial state vector as follows, using Qiskit ordering of the tensor product:

$$U|q_2 q_1\rangle|q_0\rangle = U\begin{pmatrix} x_0 \\ x_1 \\ x_2 \\ x_3 \end{pmatrix} \otimes \begin{pmatrix} 1 \\ 0 \end{pmatrix} = U\begin{pmatrix} x_0 \\ 0 \\ x_1 \\ 0 \\ x_2 \\ 0 \\ x_3 \\ 0 \end{pmatrix}$$

$$= \begin{pmatrix} 1 & 0 & 0 & 0 & 0 & 0 & 0 & 0 \\ 0 & 1 & 0 & 0 & 0 & 0 & 0 & 0 \\ 0 & 0 & \frac{1}{2} & \frac{1}{2} & \frac{1}{2} & -\frac{1}{2} & 0 & 0 \\ 0 & 0 & \frac{1}{2} & \frac{1}{2} & -\frac{1}{2} & \frac{1}{2} & 0 & 0 \\ 0 & 0 & \frac{1}{2} & -\frac{1}{2} & \frac{1}{2} & \frac{1}{2} & 0 & 0 \\ 0 & 0 & -\frac{1}{2} & \frac{1}{2} & \frac{1}{2} & \frac{1}{2} & 0 & 0 \\ 0 & 0 & 0 & 0 & 0 & 0 & 1 & 0 \\ 0 & 0 & 0 & 0 & 0 & 0 & 0 & 1 \end{pmatrix} \begin{pmatrix} x_0 \\ 0 \\ x_1 \\ 0 \\ x_2 \\ 0 \\ x_3 \\ 0 \end{pmatrix} = \begin{pmatrix} x_0 \\ 0 \\ \frac{1}{2}x_1 + \frac{1}{2}x_2 \\ \frac{1}{2}x_1 - \frac{1}{2}x_2 \\ \frac{1}{2}x_1 + \frac{1}{2}x_2 \\ -\frac{1}{2}x_1 + \frac{1}{2}x_2 \\ x_3 \\ 0 \end{pmatrix}$$

Computing the final state when the control qubit $|q_0\rangle$ is measured in state $|0\rangle$

If the control qubit $|q_0\rangle$ is measured in state $|0\rangle$, then the final state is computed by discarding all amplitudes that do not contribute to this outcome, $|001\rangle, |011\rangle, |101\rangle, |111\rangle$, and then renormalizing:

$$
P_0 \begin{pmatrix} x_0 \\ 0 \\ \frac{1}{2}x_1 + \frac{1}{2}x_2 \\ \frac{1}{2}x_1 - \frac{1}{2}x_2 \\ \frac{1}{2}x_1 + \frac{1}{2}x_2 \\ -\frac{1}{2}x_1 + \frac{1}{2}x_2 \\ x_3 \\ 0 \end{pmatrix} = \begin{pmatrix} x_0 \\ 0 \\ \frac{1}{2}x_1 + \frac{1}{2}x_2 \\ 0 \\ \frac{1}{2}x_1 + \frac{1}{2}x_2 \\ 0 \\ x_3 \\ 0 \end{pmatrix} \Big/ \sqrt{|x_0|^2 + \frac{1}{2}|x_1 + x_2|^2 + |x_3|^2}
$$

Recall that $|\Phi_+\rangle, |\Phi_-\rangle, |\Psi_+\rangle$ and $|\Psi_-\rangle$ are the Bell states we introduced in *Section 3.2.5, Creation of a Bell state.* The amplitudes x_0 of $|00\rangle$ and x_3 of $|11\rangle$ are left unchanged up to a renormalization factor in the final state. The symmetrized Bell state $|\Phi_+\rangle = \frac{1}{\sqrt{2}}(|00\rangle + |11\rangle)$ is left unchanged. The amplitudes x_1 of $|01\rangle$ and x_2 of $|10\rangle$ are mixed in the Bell state $|\Psi_+\rangle = \frac{1}{\sqrt{2}}(|01\rangle + |10\rangle)$, which is symmetrized.

Computing the final state when the control qubit $|q_0\rangle$ is measured in state $|1\rangle$

If the control qubit $|q_0\rangle$ is measured in state $|1\rangle$, then the final state is computed by discarding all amplitudes that do not contribute to this outcome, $|000\rangle, |010\rangle, |100\rangle$, and $|110\rangle$, and then renormalizing:

$$
P_1 \begin{pmatrix} x_0 \\ 0 \\ \frac{1}{2}x_1 + \frac{1}{2}x_2 \\ \frac{1}{2}x_1 - \frac{1}{2}x_2 \\ \frac{1}{2}x_1 + \frac{1}{2}x_2 \\ -\frac{1}{2}x_1 + \frac{1}{2}x_2 \\ x_3 \\ 0 \end{pmatrix} = \begin{pmatrix} 0 \\ 0 \\ 0 \\ \frac{1}{2}x_1 - \frac{1}{2}x_2 \\ 0 \\ -\frac{1}{2}x_1 + \frac{1}{2}x_2 \\ 0 \\ 0 \end{pmatrix} \Big/ \sqrt{\frac{1}{2}|x_1 - x_2|^2}
$$

The only non-null amplitudes of the final state are those in the $|01\rangle$ and $|10\rangle$ subspace, which are mixed in the Bell state $|\Psi_-\rangle = \frac{1}{\sqrt{2}}(|01\rangle - |10\rangle)$, which is antisymmetrized.

These properties of symmetry and antisymmetry are key to efficient implementations of the **Variational Quantum Eigensolver (VQE)** algorithms [Gard] that we will cover in *Chapter 6, Variational Quantum Eigensolver Algorithm*.

3.4.4. Post selecting the control qubit until the desired state is obtained

We define the `post_select()` function, which performs a loop that executes the swapper circuit and measures the state of the control qubit $|q_0\rangle$ until the desired symmetrized or antisymmetrized state is obtained, or until the maximum number of iterations is reached:

- Append a circuit created by the `swapper()` function.
- Measure the control qubit $|q_0\rangle$. If we get 0, then qubits $|q_1\rangle$ and $|q_2\rangle$ are in a symmetrized state and if we get 1, then qubits $|q_1\rangle$ and $|q_2\rangle$ are in an antisymmetrized state.

Then, `post_select()` calls the `proc_result()` function to process the results.

The `post_select()` function has the following input parameters:

- `simulator`, by default `statevector_simulator`, which simulates perfect qubits.
- `symm`: set to `True` to get a symmetrized state and `False` to get an antisymmetrized state.
- `shots` is the number of shots, and by default is set to `1`.
- `max_iter` is the maximum number of iterations, and by default is set to `20`.
- `swap_test` is set to `True` to perform a swap test to determine whether the final state is permutation symmetric or permutation asymmetric, and by default is set to `False`.

Here is the code:

```
def post_select(qc, q, c, symm=True, simulator='statevector_
simulator', shots=1, max_iter=20, swap_test=False):
   backend = Aer.get_backend(simulator)
   s = qi.Statevector.from_instruction(qc)
```

```
    display(array_to_latex(s, prefix="\\
text{Initial state} = ", precision = 2))
  done = False
  iter = 0
  while not done and iter < max_iter:
    qc.append(swapper(draw=(iter==0)), qargs=q)
    qc.measure(q[0], c[0]) # Measure control qubit q[0]
    qc.save_statevector(label=str(iter)) # Save the current
simulator state vector
    job = execute(qc, backend, shots=shots) # Execute the
Simulator
    result = job.result()
    counts = result.get_counts(qc)
    for k, v in counts.items():
      if symm and k == '0' and v > shots/2:
        done = True
      elif not symm and k == '1' and v > shots/2:
        done = True
    if not done:
      qc.reset(q[0])
      iter += 1
  success = proc_result(result, iter, counts, max_iter=max_
iter, symm=symm, simulator=simulator, swap_test=swap_test)
  return result, success
```

The proc_result() function processes the results, displays the saved state vector, and calls the factor() function. If the input parameter swap_test is set to True, it calls the swap_check() function, which tests whether the final state is permutation symmetric or permutation asymmetric.

It returns Success, a Boolean; True if the desired state has been obtained, False otherwise:

```
def proc_result(result, iter, counts, max_
iter=20, symm=True, simulator='statevector_simulator', swap_
test=False):
  if symm:
    print("Preparing a permutation symmetric state")
  else:
```

```
        print("Preparing a permutation antisymmetric state")
    print("simulator:", simulator)
    print("counts: ", counts)
    if iter >= max_iter:
        print("Post selection unsuccessful iteration {}".
format(iter))
        success = False
    else:
        print("Post selection successful iteration {}".
format(iter))
        success = True
        s = result.data()[str(iter)]
        factor(s, symm) # Call factor()
        if swap_test:
            swap_check(qc, q, iter, symm, s, simulator=simulator)
        print(" ") # Display Density matrix of the final state
        display(array_to_latex(qi.DensityMatrix(s), prefix="\\
text{Density matrix of the final state: }", precision = 2))
        display(plot_state_city(s, title='Cityscape plot of the
final state')) # Display Cityscape plot of the final state
    return success
```

The sym_test() function determines whether two amplitudes of a state vector are equal or opposite to one another and the sum of their modulus squared is equal to 1:

```
def sym_test(s, symm, i0, i1):
    if symm:
        b = np.isclose(np.abs(s[i0]-s[i1]), 0, rtol=_EPS) and np.
isclose(np.abs(s[i0]**2 + s[i1]**2), 1, rtol=1e-4)
    else:
        b = np.isclose(np.abs(s[i0]+s[i1]), 0, rtol=_EPS) and np.
isclose(np.abs(s[i0]**2 + s[i1]**2), 1, rtol=1e-4)
    return b
```

The `factor()` function attempts to factor the final state into a tensor product of the control qubit $|q_0\rangle$ and the permutation symmetric Bell states $|\Psi_+\rangle = \frac{1}{\sqrt{2}}(|01\rangle + |10\rangle)$ or $|\Phi_+\rangle = \frac{1}{\sqrt{2}}(|00\rangle + |11\rangle)$, or the permutation antisymmetric Bell states $|\Psi_-\rangle = \frac{1}{\sqrt{2}}(|01\rangle - |10\rangle)$ or $|\Phi_-\rangle = \frac{1}{\sqrt{2}}(|00\rangle - |11\rangle)$, which were introduced in *Section 3.2.5, Creation of a Bell state*:

```
def factor(s, symm):
  b0 = np.allclose(s, [1, 0, 0, 0, 0, 0, 0, 0], rtol=_EPS)
  b1 = np.allclose(s, [0, 0, 0, 0, 0, 0, 1, 0], rtol=_EPS)
  b2 = sym_test(s, symm, 2, 4)
  b3 = sym_test(s, symm, 3, 5)
  b4 = sym_test(s, symm, 0, 6)
  b5 = sym_test(s, symm, 1, 7)
  df = {b0: "|00", b1: "|11", b2: "|\\Psi_+", b3: "|\\Psi_-", b4: "|\\Phi_+", b5: "|\\Phi_-"}
  found = False
  for k, v in df.items():
    if not found and symm and k:
      display(array_to_latex([s], prefix = "\\text{Symmetrized state: }" + v + "\\rangle, \\text{ Final state: }" + v + " \\rangle |0 \\rangle =", precision = 2))
      found = True
    elif not found and not symm and k:
      display(array_to_latex([s], prefix = "\\text{Antisymmetrized state: }" + v + "\\rangle, \\text{ Final state: }" + v + " \\rangle |1 \\rangle =", precision = 2))
      found = True
  if not found:
    display(array_to_latex(s, prefix="\\text{Final state} = ", precision = 2))
  return
```

The `swap_check()` function tests whether the final state is permutation symmetric or permutation asymmetric. It calls the `swap()` function and then compares the states before and after the swap:

```
def swap_check(qc, q, iter, symm, s, simulator='statevector_simulator'):
```

```
    s21 = swap(qc, q, iter, simulator=simulator)
    if symm:
        if np.allclose(s-s21, 0, rtol=_EPS):
            print("Swap test confirms that final state is permutation
symmetric")
    else:
        if np.allclose(s+s21, 0, rtol=_EPS):
            print("Swap test confirms that final state is permutation
asymmetric")
    return
```

The swap() function performs a swap of qubits $|q_1\rangle$ and $|q_2\rangle$), measures the control qubit, and returns the final state vector for comparison purposes:

```
def swap(qc, q, iter, simulator='statevector_simulator'):
    backend = Aer.get_backend(simulator)
    qc.swap(q[1], q[2])
    qc.measure(q[0], c[0]) # Measure control qubit q[0]
    qc.save_statevector(label=str(iter+1)) # Save the current
simulator state vector
    job = execute(qc, backend, shots=1) # Execute the Simulator
    result = job.result()
    s21 = result.data()[str(iter+1)]
    return s21
```

3.4.5. Examples of final symmetrized and antisymmetrized states

We now implement five experiments creating:

- A symmetrized state from state $|10\rangle$
- An antisymmetrized state from state $|10\rangle$
- A symmetrized state from qubits initialized with random states
- An antisymmetrized state from qubits initialized with random states
- A symmetrized state from the Bell state $|\Phi_+\rangle$

Experiment creating a symmetrized state from state $|10\rangle$

We create two state vectors in states $|1\rangle$ and $|0\rangle$, we give them as input to the `setup_qc()` function, and then we call the `post_select()` function with `symm` set to `True`:

```
s1 = qi.Statevector([0, 1])
s2 = qi.Statevector([1, 0])
qc, q, c = setup_qc(s1, s2)
result, success = post_select(qc, q, c, symm=True)
```

We have obtained the symmetrized state $|\Psi_+\rangle = \dfrac{1}{\sqrt{2}}(|01\rangle + |10\rangle)$, as shown in *Figure 3.24*:

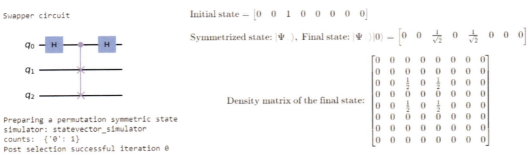

Figure 3.24 – Symmetrized state obtained from state $|10\rangle$

We show the density matrix using a cityscape plot of the final state in *Figure 3.25*:

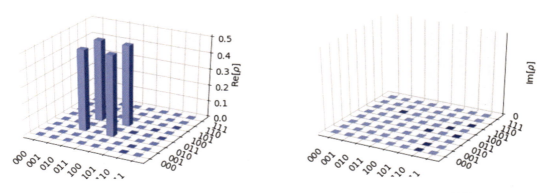

Figure 3.25 – Cityscape plot of the symmetrized state obtained from state $|\mathbf{10}\rangle$

We display the quantum circuit with the `draw()` method of the quantum circuit class:

```
qc.draw(output='mpl', plot_barriers=False)
```

After one iteration, the quantum circuit looks like *Figure 3.26*:

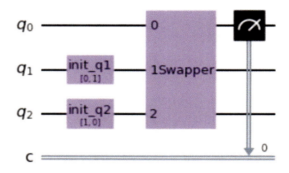

Figure 3.26 – Quantum circuit after one iteration

Experiment creating an antisymmetrized state from state |10⟩

We create two state vectors in states $|1\rangle$ and $|0\rangle$, we give them as input to the `setup_qc()` function, and then we call the `post_select()` function with `symm` set to `False`:

```
s1 = qi.Statevector([0, 1])
s2 = qi.Statevector([1, 0])
qc, q, c = setup_qc(s1, s2)
result, success = post_select(qc, q, c, symm=False)
```

We have obtained the antisymmetrized state $|\Psi_-\rangle = \frac{1}{\sqrt{2}}(|01\rangle - |10\rangle)$, as shown in *Figure 3.27*:

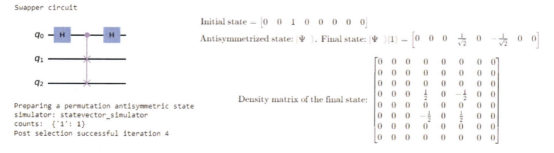

Figure 3.27 – Antisymmetrized state obtained from state |10⟩

We show the density matrix using a cityscape plot of the final state in *Figure 3.28*:

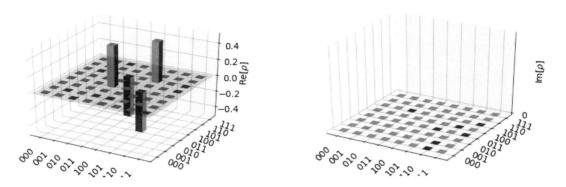

Figure 3.28 – Cityscape plot of the antisymmetrized state obtained from state |10⟩

Experiment creating a symmetrized state from qubits initialized with random states

We create two state vectors in random states with `init_random()`, we give them as input to the `setup_qc()` function, and then we call the `post_select()` function with `symm` set to `True` and `swap_test` set to `True` to confirm that the final state is indeed permutation symmetric. We expect the final state to have four non-null amplitudes, based on the computation made in *Section 3.4.3, Creating a circuit that swaps two qubits with a controlled swap gate, Computing the final state when the control qubit |q₀⟩ is measured in state |0⟩*:

```
s1, s2 = init_random()
qc, q, c = setup_qc(s1, s2)
result, success = post_select(qc, q, c, symm=True, swap_
test=True)
```

In *Figure 3.29*, we show a result where the final state has four non-null amplitudes, as expected, and the swap test has confirmed that the final state is permutation symmetric:

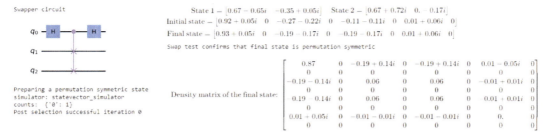

Figure 3.29 – Symmetrized state obtained from qubits initialized in random states

In the cityscape plot of the final state shown in *Figure 3.30*, we see that the state $|000\rangle$ has the largest probability to come out after measuring the final state:

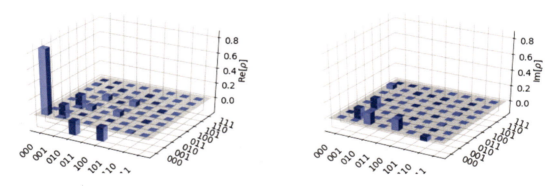

Figure 3.30 – Cityscape plot of symmetrized state obtained from qubits initialized in random states

Experiment creating an antisymmetrized state from qubits initialized with random states

We create two state vectors in random states with `init_random()`, we give them as input to the `setup_qc()` function, and then we call the `post_select()` function with `symm` set to `False`:

```
s1, s2 = init_random()
qc, q, c = setup_qc(s1, s2)
result, success = post_select(qc, q, c, symm=False)
```

We have obtained the antisymmetrized state $|\Psi_-\rangle = \frac{1}{\sqrt{2}}(|01\rangle - |10\rangle)$ up to a global phase, as shown in *Figure 3.31*:

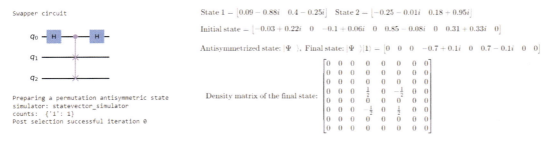

Figure 3.31 – Antisymmetrized state obtained from qubits initialized in random states

We show the density matrix using a cityscape plot of the final state in *Figure 3.32*:

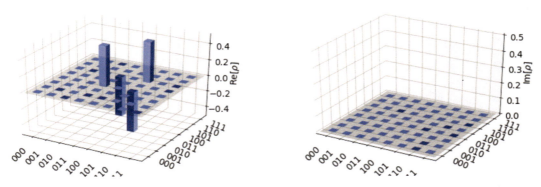

Figure 3.32 – Cityscape plot of the antisymmetrized state obtained from qubits initialized in random states

Experiment creating a symmetrized state from the Bell state $|\Phi_+\rangle$

We define a function called `setup1_qc()`, which sets up a quantum circuit that puts qubits $|q_1\rangle$ and $|q_2\rangle$ into the Bell state $|\Phi_+\rangle = \frac{1}{\sqrt{2}}(|00\rangle + |11\rangle)$, as follows:

```
def setup1_qc(draw=False):
    q = QuantumRegister(3, 'q') # register of 3 qubits
    c = ClassicalRegister(1, name="c") # and 1 classical register
    qc = QuantumCircuit(q,c)
    qc.h(q[1])
    qc.cx(q[1], q[2])
```

```
    qc.barrier()
    if draw:
        display(qc.draw(output='mpl'))
    return qc, q, c
```

We execute the following code, which calls `setup1_qc()` to create a Bell state $|\Phi_+\rangle$ and then calls the `post_select()` function with `symm` set to `True`:

```
qc, q, c = setup1_qc()
result, success = post_select(qc, q, c, symm=True)
```

As expected in *Section 3.4.3, Creating a circuit that swaps two qubits with a controlled swap gate*, the swapper circuit followed by a measurement of the qubit $|q_0\rangle$ in state 0 leaves the symmetrized Bell state $|\Phi_+\rangle = \frac{1}{\sqrt{2}}(|00\rangle + |11\rangle)$ unchanged, as shown in *Figure 3.33*:

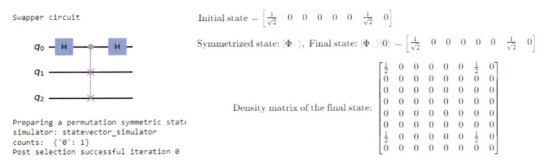

Figure 3.33 – Symmetrized state obtained from the Bell state $|\Phi_+\rangle$

We show the density matrix using a cityscape plot of the final state in *Figure 3.34*:

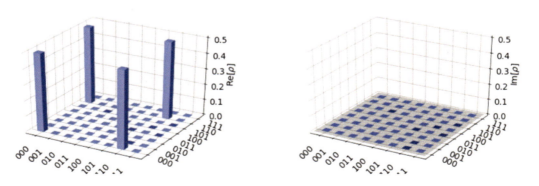

Figure 3.34 – Cityscape plot of the final state obtained from the Bell state $|\Phi_+\rangle$

References

[ChemChiral] 5.1 Chiral Molecules, Chemistry LibreTexts, 5 Jul 2015, `https://chem.libretexts.org/Bookshelves/Organic_Chemistry/Map%3A_Organic_Chemistry_(Vollhardt_and_Schore)/05._Stereoisomers/5.1%3A_Chiral__Molecules`

[Corcoles] A. D. Córcoles, Maika Takita, Ken Inoue, Scott Lekuch, Zlatko K. Minev, Jerry M. Chow, and Jay M. Gambetta, Exploiting Dynamic Quantum Circuits in a Quantum Algorithm with Superconducting Qubits, Phys. Rev. Lett. 127, 100501, 31 August 2021, `https://journals.aps.org/prl/abstract/10.1103/PhysRevLett.127.100501`

[Crockett] Christopher Crockett, Superpositions of Chiral Molecules, September 14, 2021, Physics 14, s108, `https://physics.aps.org/articles/v14/s108`

[Gard] Gard, B.T., Zhu, L., Barron, G.S. et al., Efficient symmetry-preserving state preparation circuits for the variational quantum eigensolver algorithm, npj Quantum Inf 6, 10 (2020), `https://doi.org/10.1038/s41534-019-0240-1`

[Grok] Grok the Bloch Sphere, `https://javafxpert.github.io/grok-bloch/`

[IBM_CEO] IBM CEO: Quantum computing will take off 'like a rocket ship' this decade, Fast Company, Sept 28, 2021., `https://www.fastcompany.com/90680174/ibm-ceo-quantum-computing-will-take-off-like-a-rocket-ship-this-decade`

[IBM_comp1] Welcome to IBM Quantum Composer, `https://quantum-computing.ibm.com/composer/docs/iqx/`

[IBM_comp2] IBM Quantum Composer, `https://quantum-computing.ibm.com/composer/files/new`

[IBM_mid] Mid-Circuit Measurements Tutorial, IBM Quantum systems, `https://quantum-computing.ibm.com/lab/docs/iql/manage/systems/midcircuit-measurement/`

[NumPy] NumPy: the absolute basics for beginners, `https://numpy.org/doc/stable/user/absolute_beginners.html`

[Qiskit] Qiskit, `https://qiskit.org/`

[QuTiP] QuTiP, Plotting on the Bloch Sphere, `https://qutip.org/docs/latest/guide/guide-bloch.html`

[Rioux] Mach-Zehnder Polarizing Interferometer Analyzed Using Tensor Algebra, `https://faculty.csbsju.edu/frioux/photon/MZ-Polarization.pdf`

[Spheres] How to Prepare a Permutation Symmetric Multiqubit State on an Actual Quantum Computer, https://spheres.readthedocs.io/en/stable/notebooks/9_Symmetrized_Qubits.html

[Stickler] B. A. Stickler et al., Enantiomer superpositions from matter-wave interference of chiral molecules, Phys. Rev. X 11, 031056 (2021), https://journals.aps.org/prx/abstract/10.1103/PhysRevX.11.031056

[Wonders] Optical Isomers, Enantiomers and Chiral Molecules, WondersofChemistry, https://www.youtube.com/watch?v=8TIZdWR4gIU

4
Molecular Hamiltonians

"The best that most of us can hope to achieve in physics is simply to misunderstand at a deeper level."

– *Wolfgang Pauli*

Figure 4.1 – Wolfgang Pauli reaching for a deeper understanding of the antisymmetry related to fermionic spin [authors]

At the end of Wolfgang Pauli's 1946 Nobel lecture [Pauli] he states:

"I may express my critical opinion, that a correct theory should neither lead to infinite zero-point energies nor to infinite zero charges, that it should not use mathematical tricks to subtract infinities or singularities, nor should it invent a hypothetical world which is only a mathematical fiction before it is able to formulate the correct interpretation of the actual world of physics."

The concepts in this chapter have a mathematical formulation and do not have a physical or chemical reality. In other words, there are standard approximations used that allow the determination of useful chemical quantities. The calculations that use the approximations do not represent exact quantities; rather, they are approximate quantities. Therefore, these approximations require a deeper understanding in order to obtain refinement and better answers.

Furthermore, the approximations used in this chapter apply only to fermionic (electronic) systems. The extension to bosonic systems is an area of research by the broader scientific community. We will see the implementation and use of virtual orbitals and both occupied and unoccupied orbitals in calculations of ground state energy.

We mention different levels of implementation of the **Hartree-Fock (HF)** theory: **Restricted Hartree-Fock (RHF)**, **Restricted Open-shell Hartree-Fock (ROHF)**, and **Unrestricted Hartree-Fock (UHF)**; however, through Qiskit, we will only show RHF. There are post-HF methodologies that one can use, such as **Coupled-Cluster (CC)**, which we will be using in *Chapter 5, Variational Quantum Eigensolver*.

In this chapter, we solve the fermionic Hamiltonian equation for a hydrogen molecule and a lithium hydride molecule and will cover the following topics:

- *Section 4.1, Born-Oppenheimer approximation*
- *Section 4.2, Fock space*
- *Section 4.3, Fermionic creation and annihilation operators*
- *Section 4.4, Molecular Hamiltonian in the Hartree-Fock orbitals basis*
- *Section 4.5, Basis sets*
- *Section 4.6, Constructing a fermionic Hamiltonian with Qiskit Nature*
- *Section 4.7, Fermion to qubit mappings*
- *Section 4.8, Constructing a qubit Hamiltonian operator with Qiskit Nature*

Technical requirements

We provide a link to a companion Jupyter notebook of this chapter, which has been tested in the Google Colab environment, which is free and runs entirely in the cloud, and in the IBM Quantum Lab environment. Please refer to *Appendix B – Leveraging Jupyter Notebooks in the Cloud*, for more information. The companion Jupyter notebook automatically installs the following list of libraries:

- NumPy [NumPy], an open source Python library that is used in almost every field of science and engineering

- Qiskit [Qiskit], an open source SDK for working with quantum computers at the level of pulses, circuits, and application modules

- Qiskit visualization support to enable the use of visualizations and Jupyter notebooks

- Qiskit Nature [Qiskit_Nature] [Qiskit_Nat_0], a unique platform that bridges the gap between natural sciences and quantum simulations

- **Python-based Simulations of Chemistry Framework (PySCF)**, [PySCF], an open source collection of electronic structure modules powered by Python

Installing NumPy, Qiskit, and importing the various modules

Install NumPy with the following command:

```
pip install numpy
```

Install Qiskit with the following command:

```
pip install qiskit
```

Install Qiskit visualization support with the following command:

```
pip install 'qiskit[visualization]'
```

Install Qiskit Nature with the following command:

```
pip install qiskit-nature
```

Install PySCF with the following command:

```
pip install pyscf
```

Import NumPy with the following command:

```
import numpy as np
```

Import Matplotlib, a comprehensive library for creating static, animated, and interactive visualizations in Python with the following command:

```
import matplotlib.pyplot as plt
```

Import the required functions and class methods with the following commands. The `array_to_latex function()` returns a LaTeX representation of a complex array with dimension 1 or 2:

```
from qiskit.visualization import array_to_latex, plot_bloch_
vector, plot_bloch_multivector, plot_state_qsphere, plot_state_
city
```

```
from qiskit import QuantumRegister, ClassicalRegister,
QuantumCircuit, transpile
```

```
from qiskit import execute, Aer
```

```
import qiskit.quantum_info as qi
```

```
from qiskit.extensions import Initialize
```

```
from qiskit.providers.
aer import extensions   # import aer snapshot instructions
```

Import Qiskit Nature libraries with the following commands:

```
from qiskit import Aer
```

```
from qiskit_nature.drivers import UnitsType, Molecule
```

```
from qiskit_nature.drivers.second_quantization import
ElectronicStructureDriverType,
 ElectronicStructureMoleculeDriver
```

```
from qiskit_nature.problems.second_quantization import
ElectronicStructureProblem
```

```
from qiskit_nature.mappers.second_
quantization import ParityMapper, JordanWignerMapper,
 BravyiKitaevMapper
```

```
from qiskit_nature.converters.second_quantization import
QubitConverter
```

```
from qiskit_nature.transformers.second_quantization.
electronic import ActiveSpaceTransformer, FreezeCoreTransformer
```

```
from qiskit_nature.operators.second_quantization import
FermionicOp
```

Import the Qiskit Nature property framework with the following command:

```
from qiskit_nature.properties import Property, GroupedProperty
```

Import the `ElectronicEnergy` property with the following command:

```
# https://qiskit.org/documentation/nature/tutorials/08_
property_framework.html
from qiskit_nature.properties.second_quantization.
electronic import (
    ElectronicEnergy,
    ElectronicDipoleMoment,
    ParticleNumber,
    AngularMomentum,
    Magnetization,
)
```

Import the `ElectronicIntegrals` property with the following command:

```
from qiskit_nature.properties.second_quantization.electronic.
integrals import (
    ElectronicIntegrals,
    OneBodyElectronicIntegrals,
    TwoBodyElectronicIntegrals,
    IntegralProperty,
)
from qiskit_nature.properties.second_quantization.electronic.
bases import ElectronicBasis
```

Import the math libraries with the following command:

```
import cmath
import math
```

4.1. Born-Oppenheimer approximation

Recall that the atomic orbital of an electron in an atom and the molecular orbital of an electron in a molecule are time-independent stationary states. In *Section 2.4*, *Postulate 4 – Time-independent stationary states*, we introduced the time-independent Schrödinger equation:

$$\widehat{H}|\psi\rangle = E|\psi\rangle$$

where $\widehat{H}$ is the non-relative Hamiltonian operator obtained by quantizing the classical energy in Hamilton form (first quantization), and it represents the total energy (E) of all its particles; N electrons and M nuclei. For a molecular system, the electric charge of two nuclei A and B are Z_A and Z_B with masses M_A and M_B. The position of the particles in the molecule is determined by using a **laboratory (LAB)** frame coordinate system, as shown in *Figure 4.2*, where the origin of the coordinate system is outside the molecule. The origin of the coordinate system can be placed anywhere in free space.

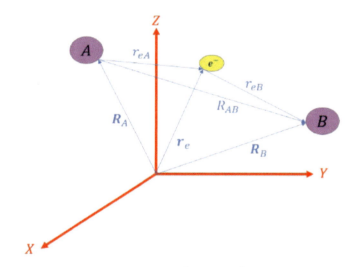

Figure 4.2 – LAB frame coordinates

The Hamiltonian in the LAB frame coordinates $\widehat{H}_{LAB}$ is:

$$\widehat{H}_{LAB} = -\frac{1}{2}\sum_{p=1}^{N}\nabla_{r_p}^2 - \sum_{A=1}^{M}\frac{1}{2M_A}\nabla_{R_A}^2 - \sum_{p=1}^{N}\sum_{A=1}^{M}\frac{Z_A}{r_{pA}} + \sum_{q>p=1}^{N}\frac{1}{r_{pq}} + \sum_{B>A=1}^{M}\frac{Z_A Z_B}{R_{AB}}$$

where in atomic units, the mass of the electron, the reduced Planck constant ($\hbar$), and the electric charge (e) are set to the value 1. The LAB Hamiltonian comprises the sum of the kinetic energy of all particles and the potential energy between all particles with the following definitions:

- $\nabla^2_{r_p}$ and $\nabla^2_{R_A}$ are the second derivative operator with respect to the position coordinates for electrons and nuclei, that is, $\nabla^2_{R_A} = \frac{\partial^2}{\partial^2_{x_A}} + \frac{\partial^2}{\partial^2_{y_A}} + \frac{\partial^2}{\partial^2_{z_A}}$, and likewise for the pth electron.

- $r_{pq} = |\mathbf{r}_p - \mathbf{r}_q|$, $r_{pA} = |\mathbf{r}_p - \mathbf{R}_A|$, and $R_{AB} = |\mathbf{R}_A - \mathbf{R}_B|$ are the distances between electrons p and q, electron p and nucleus A, and nuclei A and B determined by the Euclidean norm.

For clarity, we list the operators of the LAB Hamiltonian in *Figure 4.3*:

Operator	Description
$-\dfrac{1}{2}\sum\limits_{p=1}^{N} \nabla^2_{r_p}$	Electronic kinetic energy operator.
$-\sum\limits_{A=1}^{M} \dfrac{1}{2M_A} \nabla^2_{R_A}$	Nuclear kinetic energy operator.
$-\sum\limits_{p=1}^{N}\sum\limits_{A=1}^{M} \dfrac{Z_A}{r_{pA}}$	Potential energy between the electrons and nuclei. It is the sum of all electron-nucleus Coulomb interactions.
$\sum\limits_{q>p=1}^{n} \dfrac{1}{r_{pq}}$	Potential energy operator arising from electron-electron Coulomb repulsions.
$\sum\limits_{B>A=1}^{M} \dfrac{Z_A Z_B}{R_{AB}}$	Potential nuclear-nuclear repulsion energy operator, the sum of all nucleus-nucleus Coulomb repulsions.

Figure 4.3 – Terms of the Hamiltonian operator for a molecule

In the LAB Hamiltonian, the energy of the molecular system is continuous, not discrete. The **center-of-mass (COM)** motion does not yield any change to the energy of the internal states of the system and can be factored out. The internal states are quantized and invariant to translations. These states are not affected by translational and rotational motions in free space. The nuclei can still move around the COM through vibrations and internal rotations.

In the **Born-Oppenheimer** (**BO**) approximation, we assume that the motions of the nuclei are uncoupled from the motions of the electrons, that is, a product of nuclear equations (rotational and vibrational) and electronic equations:

$$|\Psi_{total}(\boldsymbol{r}, \boldsymbol{s}, \boldsymbol{R})\rangle = \Psi_{rotational}(\boldsymbol{R})\Psi_{vibrational}(\boldsymbol{R})\Psi_{elec}(\boldsymbol{r}, \boldsymbol{s}; \boldsymbol{R})$$

where $\boldsymbol{R} = \{\boldsymbol{R}_A, \boldsymbol{R}_B, \dots, \boldsymbol{R}_M\}$ are the nuclear coordinates, $\boldsymbol{r} = \{\boldsymbol{r}_p, \boldsymbol{r}_i, \boldsymbol{r}_j, \dots, \boldsymbol{r}_N\}$ are the electron coordinates, $\boldsymbol{s} = \{\boldsymbol{s}_p, \boldsymbol{s}_i, \boldsymbol{s}_j, \dots, \boldsymbol{s}_N\}$ are the spin coordinates, and the electronic wave function ($\Psi_{elec}(\boldsymbol{r}, \boldsymbol{s}; \boldsymbol{R})$) is the condition on the nuclear coordinates ($\boldsymbol{R}$).

In the BO approximation, solving for only the electronic equation with the fixed position of the nuclei can be iterated to account for the vibrations and internal rotations of the nuclei. For each iteration, the nuclei of the atoms are fixed in space and can be thought of as a violation of the Heisenberg uncertainty principle introduced in Section 1.4., *Light and energy*. The more you know exactly where a particle is, the less you know about its momentum. In general, the internal coordinate system can be placed at the heaviest atom in the molecule or at the COM.

We are only dealing with electrons moving around stationary nuclei. Hence, if we ignore the kinetic energy coupling terms of the nuclei and the nuclei with electrons, the general electronic molecular Hamiltonian is a sum of four operators, as shown:

$$\widehat{H}_{elec} = -\frac{1}{2}\sum_{p=1}^{N}\nabla_{\boldsymbol{r}_p}^2 - \sum_{p=1}^{N}\sum_{A=1}^{M}\frac{Z_A}{r_{pA}} + \sum_{q>p=1}^{N}\frac{1}{r_{pq}} + \sum_{B>A=1}^{M}\frac{Z_A Z_B}{R_{AB}}$$

The last term, the potential nuclear repulsion energy, is not computed and is approximated with **pseudopotentials** and experimental data, and we will show this in *Figure 4.4*. This approximation has limitations that we cover further in *Chapter 6, Beyond Born-Oppenheimer*, specifically through determining potential energy surfaces.

4.2. Fock space

The electronic wave function $\Psi_{elec}(\boldsymbol{r}, \boldsymbol{s}; \boldsymbol{R})$ includes the spatial position for each electron ($\boldsymbol{r}_p = \{x_p, y_p, z_p\}$) conditioned on the nuclear positions ($\boldsymbol{R}_A = \{X_A, Y_A, Z_A\}$) and the spin directional coordinates for each electron ($\boldsymbol{s}_p = \{s_{x_p}, s_{y_p}, s_{z_p}\}$). The electronic wave function must be antisymmetric with respect to the exchange of pair particle labels because these particles are fermions. Please recall that this is the Pauli exclusion principle introduced in *Section 2.1.3, General formulation of the Pauli exclusion principle*.

Now, let's consider two electrons i and j in states $|r_i, s_i\rangle$ and $|r_j, s_j\rangle$ where the corresponding electronic wave function ($\Psi_{elec}(r_i, s_i, r_j, s_j; R)$) representing the states of the two-electron system is antisymmetric ($-$) with respect to the exchange of identical pair particles ($\hat{A}_{ij}$):

$$\hat{A}_{ij}\Psi_{elec}(r_i, s_i, r_j, s_j; R) = -\Psi_{elec}(r_i, s_i, r_j, s_j; R)$$

$$\hat{A}_{ij}\Psi_{elec}(r_i, s_i, r_j, s_j; R) = \hat{A}_{ij}\frac{1}{\sqrt{2}}\left(|r_i, s_i\rangle \otimes |r_j, s_j\rangle - |r_j, s_j\rangle \otimes |r_i, s_i\rangle\right)$$

$$= \frac{1}{\sqrt{2}}\left(|r_j, s_j\rangle \otimes |r_i, s_i\rangle - |r_i, s_i\rangle \otimes |r_j, s_j\rangle\right)$$

$$= -\frac{1}{\sqrt{2}}\left(|r_i, s_i\rangle \otimes |r_j, s_j\rangle - |r_j, s_j\rangle \otimes |r_i, s_i\rangle\right) = -\Psi_{elec}(r_i, s_i, r_j, s_j; R)$$

Since a one-body electronic wave function as presented in Section 2.1, *Postulate 1 – Wave functions*, is a product of spatial ($\psi(r_1)$) and spin components ($\chi(s_1)$), there are two scenarios where antisymmetry can be achieved:

- The spin function must be antisymmetric while the spatial is symmetric.

- The spatial function must be antisymmetric while the spin function is symmetric.

This means that when two particles are swapped, the total molecular wave function ($\Psi_{total}(r, s, R)$) describing the system also changes sign:

$$\hat{A}_{ij}\Psi_{total}(r, s, R) = -\Psi_{total}(r, s, R)$$

For N electrons in the field of nuclei, the total wave function $\Psi_{total}(r, s, R)$ can be written as a product of atomic one-body spin orbitals:

$$\Psi_{elec}(r, s; R) = \Psi_1(r, s; R) \wedge \Psi_2(r, s; R) \wedge ... \wedge \Psi_N(r, s; R)$$

and is made to be anti-symmetric through an antisymmetric tensor product ($\wedge$) [Toulouse] and through the superposition of states using the **Slater determinant** first introduced by Dirac [Kaplan]:

$$\Psi_{elec}(r, s; R) = \frac{1}{\sqrt{N!}}\begin{vmatrix} \Psi_1(r_1, s_1; R) & \Psi_1(r_2, s_2; R) & ... & \Psi_1(r_N, s_N; R) \\ \Psi_2(r_1, s_1; R) & \Psi_2(r_2, s_2; R) & ... & \Psi_2(r_N, s_N; R) \\ ... & ... & ... & ... \\ \Psi_N(r_1, s_1; R) & \Psi_N(r_2, s_2; R) & ... & \Psi_N(r_N, s_N; R) \end{vmatrix}$$

The Slater determinant wave function is antisymmetric with respect to the exchange of two electrons (permutation of two rows) or with respect to the exchange of two spin orbitals (permutation of two columns).

For two electrons ($N = 2$), the Slater determinant has the form:

$$\Psi_{elec}(\boldsymbol{r}_1, \boldsymbol{r}_2; \boldsymbol{R}) = \frac{1}{\sqrt{2}} \begin{vmatrix} \Psi_1(\boldsymbol{r}_1, \boldsymbol{s}_1; \boldsymbol{R}) & \Psi_1(\boldsymbol{r}_2, \boldsymbol{s}_2; \boldsymbol{R}) \\ \Psi_2(\boldsymbol{r}_1, \boldsymbol{s}_1; \boldsymbol{R}) & \Psi_2(\boldsymbol{r}_2, \boldsymbol{s}_2; \boldsymbol{R}) \end{vmatrix}$$

$$= \frac{1}{\sqrt{2}} \left(\Psi_1(\boldsymbol{r}_1, \boldsymbol{s}_1; \boldsymbol{R}) \otimes \Psi_2(\boldsymbol{r}_2, \boldsymbol{s}_2; \boldsymbol{R}) - \Psi_1(\boldsymbol{r}_2, \boldsymbol{s}_2; \boldsymbol{R}) \otimes \Psi_2(\boldsymbol{r}_1, \boldsymbol{s}_1; \boldsymbol{R}) \right)$$

For three electrons ($N = 3$), the Slater determinant has the form:

$$\Psi_{elec}(\boldsymbol{r}_1, \boldsymbol{r}_2, \boldsymbol{r}_3, \boldsymbol{s}_1, \boldsymbol{s}_2, \boldsymbol{s}_3; \boldsymbol{R}) = \frac{1}{\sqrt{6}} \begin{vmatrix} \Psi_1(\boldsymbol{r}_1, \boldsymbol{s}_1; \boldsymbol{R}) & \Psi_1(\boldsymbol{r}_2, \boldsymbol{s}_2; \boldsymbol{R}) & \Psi_1(\boldsymbol{r}_3, \boldsymbol{s}_3; \boldsymbol{R}) \\ \Psi_2(\boldsymbol{r}_1, \boldsymbol{s}_1; \boldsymbol{R}) & \Psi_2(\boldsymbol{r}_2, \boldsymbol{s}_2; \boldsymbol{R}) & \Psi_2(\boldsymbol{r}_3, \boldsymbol{s}_3; \boldsymbol{R}) \\ \Psi_3(\boldsymbol{r}_1, \boldsymbol{s}_1; \boldsymbol{R}) & \Psi_3(\boldsymbol{r}_2, \boldsymbol{s}_2; \boldsymbol{R}) & \Psi_3(\boldsymbol{r}_3, \boldsymbol{s}_3; \boldsymbol{R}) \end{vmatrix}$$

$$= \frac{1}{\sqrt{6}} \{ \Psi_1(\boldsymbol{r}_1, \boldsymbol{s}_1; \boldsymbol{R}) \otimes \left(\Psi_2(\boldsymbol{r}_2, \boldsymbol{s}_2; \boldsymbol{R}) \otimes \Psi_3(\boldsymbol{r}_3, \boldsymbol{s}_3; \boldsymbol{R}) - \Psi_2(\boldsymbol{r}_3, \boldsymbol{s}_3; \boldsymbol{R}) \otimes \Psi_3(\boldsymbol{r}_2, \boldsymbol{s}_2; \boldsymbol{R}) \right)$$

$$- \Psi_1(\boldsymbol{r}_2, \boldsymbol{s}_2; \boldsymbol{R}) \otimes \left(\Psi_2(\boldsymbol{r}_1, \boldsymbol{s}_1; \boldsymbol{R}) \otimes \Psi_3(\boldsymbol{r}_3, \boldsymbol{s}_3; \boldsymbol{R}) - \Psi_2(\boldsymbol{r}_3, \boldsymbol{s}_3; \boldsymbol{R}) \otimes \Psi_3(\boldsymbol{r}_1, \boldsymbol{s}_1; \boldsymbol{R}) \right)$$

$$+ \Psi_1(\boldsymbol{r}_3, \boldsymbol{s}_3; \boldsymbol{R}) \otimes \left(\Psi_2(\boldsymbol{r}_1, \boldsymbol{s}_1; \boldsymbol{R}) \otimes \Psi_3(\boldsymbol{r}_2, \boldsymbol{s}_2; \boldsymbol{R}) - \Psi_2(\boldsymbol{r}_2, \boldsymbol{s}_2; \boldsymbol{R}) \otimes \Psi_3(\boldsymbol{r}_1, \boldsymbol{s}_1; \boldsymbol{R}) \right) \}$$

The **Fock space** [Fock] is the Hilbert space in which the Slater determinant wave functions belong. By definition, a Fock space is the sum of a set of Hilbert spaces representing at least three important configurations:

- The zero-particle state also called the vacuum state is interpreted as the absence of an electron in an orbital: $|\text{vac}\rangle$ or $|0\rangle$.

- One-particle states: $|\boldsymbol{r}_\text{p}, \boldsymbol{s}_\text{p}\rangle$ or $|1\rangle$.

- Two-particle states: $|\boldsymbol{r}_i, \boldsymbol{s}_i, \boldsymbol{r}_j, \boldsymbol{s}_j\rangle = \frac{1}{\sqrt{2}} \left(|\boldsymbol{r}_j, \boldsymbol{s}_j\rangle \otimes |\boldsymbol{r}_i, \boldsymbol{s}_i\rangle - v |\boldsymbol{r}_i, \boldsymbol{s}_i\rangle \otimes |\boldsymbol{r}_j, \boldsymbol{s}_j\rangle \right)$ or $|11\rangle$.

The number of states of an n-particle subspace of the Fock space of N electrons is:

$$\binom{N}{n} = \frac{N!}{n! \, (N - n)}$$

where $\binom{N}{n}$ denotes the number of n-combinations from a set of N elements.

If we have 5 electrons with a 3-particle subspace:

$$\binom{5}{3} = \frac{5!}{3! \, (5 - 3)} = \frac{5 \times 4 \times 3 \times 2 \times 1}{3 \times 2 \times 1 \times 2} = 10$$

The total number of states in a Fock space of N electrons is [Wiki-Comb]:

$$\sum_{n=0}^{N} \binom{N}{n} = \sum_{n=0}^{N} \frac{N!}{n! \, (N - n)} = 2^N$$

What we present here in this section is only a mathematical construction and does not represent a physical reality nor a chemical actuality. Therefore, in some ways, it is difficult to relate to the ideas and terms we use to actual chemistry taking place. However, the Fock space is exploited in quantum computing because there is a one-to-one mapping between the electron space of a molecule and the qubit space; but it is not a necessary mapping. There are other mappings that are more computationally advantageous, such as the one presented in *Section 4.7.4, Bravyi-Kitaev transformation*.

4.3. Fermionic creation and annihilation operators

In the previous section, we mentioned that the Fock space is a mathematical construction and does not represent a physical reality nor a chemical actuality. However, please keep in mind that in a molecule, each electron can occupy only one spin-orbit at a time and no two electrons can occupy the same spin-orbit.

Now we further consider a subspace of the Fock space, which is spanned by the **occupation number** of the spin-orbits, which is described by 2^N electronic basis states $|f_0 \ldots f_{N-1}\rangle$, where $f_j \in \{0,1\}$ is the occupation number of orbital j.

The spin-orbital state j not occupied by an electron is represented by $|f_0 \ldots 0_j \ldots f_{N-1}\rangle$.

We define a set of fermionic annihilation operators $\{\hat{a}_i\}_{i=0}^{N-1}$ and creation operators $\{\hat{a}_j^\dagger\}_{j=0}^{N-1}$, which act on local electron modes, and which satisfy the following **anti-commutation** relations:

$$\{\hat{a}_i, \hat{a}_j^\dagger\} = \hat{a}_i^\dagger \hat{a}_j + \hat{a}_j \hat{a}_i^\dagger = \delta_{ij} = \begin{cases} 0, & i \neq j \\ 1, & i = j \end{cases}$$

$$\{a_i^\dagger, a_j^\dagger\} = \{\hat{a}_i, \hat{a}_j\} = 0$$

where δ_{ji} is the Dirac delta function. The operators $\{\hat{a}_j^\dagger \hat{a}_j\}_{j=0}^{n-1}$ are called the **occupation number operators** and commute with one another.

A **fermionic operator** is a linear combination of products of **creation** and **annihilation operators**, which we discuss next.

4.3.1. Fermion creation operator

The fermionic creation operator $\hat{a}_i^\dagger$ raises by one unit the number of particles sitting in the ith fermionic orbital:

$$\hat{a}_i^\dagger |\dots m_i \dots\rangle = (1 - m_i)(-1)^{\Sigma_{j<i} m_j} |\dots (m+1)_i \dots\rangle$$

where:

- m_i and $(m+1)_i$ are the number of particles sitting in the ith fermionic orbital.

- $(1 - m_i)$ is a pre-factor that annihilates the state if we had an electron in the ith fermionic orbital, that is, if $m_i = 1$.

- The phase factor $(-1)^{\Sigma_{j<i} m_j}$ keeps the anti-symmetric properties of the whole superposition of states.

4.3.2. Fermion annihilation operator

The fermion annihilation operator $\hat{a}_i$ lowers by one unit the number of particles sitting in the ith fermionic orbital:

$$\hat{a}_i |\dots m_i \dots\rangle = m_i (-1)^{\Sigma_{j<i} m_j} |\dots (m-1)_i \dots\rangle$$

where:

- m_i and $(m-1)_i$ are the number of particles sitting in the ith fermionic orbital.

- m_i is a pre-factor that annihilates the state in the Slater determinant if there is no electron in the ith fermionic orbital, that is, if $m_i = 0$.

- The phase factor $(-1)^{\Sigma_{j<i} m_j}$ keeps the anti-symmetric properties of the whole superposition of states.

We now see how to write the electronic molecular Hamiltonian as a linear combination of products of creation and annihilation operators.

4.4. Molecular Hamiltonian in the Hartree-Fock orbitals basis

For mapping the original **electronic structure Hamiltonian** into the corresponding **qubit Hamiltonian**, we work in the second quantization formalism of quantum mechanics. Recall we introduced the first quantization in *Section 4.1, Born-Oppenheimer approximation.*

The **Hartree-Fock (HF)** method approximates an N-body problem into N one-body problems where each electron evolves in the **mean field** of the other electrons.

We can rewrite the electronic molecular Hamiltonian ($\widehat{H}_{elec}$) as a linear combination of products of creation and annihilation operators (summarized in *Figure 4.4*):

$$\widehat{H}_{elec} = \sum_{i,j} h_{i,j}\, \hat{a}_i^\dagger \hat{a}_j + \frac{1}{2} \sum_{i,j,k,l} g_{i,j,k,l}\, \hat{a}_i^\dagger \hat{a}_j^\dagger \hat{a}_l \hat{a}_k + E_{NN}$$

where $\hat{a}_j$ removes an electron from spin-orbital j, and $\hat{a}_i^\dagger$ creates an electron in spin-orbital i. The operation $\hat{a}_i^\dagger \hat{a}_j$ is the **excitation operator**, which excites an electron from the **occupied spin-orbital** $\psi_j(r_p)\chi_j(s_p)$ into the **unoccupied orbital** $\psi_i(r_p)\chi_i(s_p)$. Constructing these were introduced in *Section 2.1, Postulate 1 – Wave functions.* The nuclear-nuclear (NN) repulsion energy (E_{NN}) is approximated by pseudopotentials and experimental data as mentioned in Section 4.1, *Born-Oppenheimer approximation.*

Molecular Hamiltonian in the BO approximation	Linear combination of products of creation and annihilation operators	Description
$-\dfrac{1}{2}\sum_{p=1}^{N} \nabla_{r_p}^2$	$-\dfrac{1}{2}\sum_{i,j} \langle i|\nabla_i^2|j\rangle\, \hat{a}_i^\dagger \hat{a}_j$	Electronic kinetic energy operator.
$-\sum_{p=1}^{N}\sum_{A=1}^{M} \dfrac{Z_A}{r_{pA}}$	$\sum_{i,j} \langle i|\dfrac{Z_A}{r_{iA}}|j\rangle\, \hat{a}_i^\dagger \hat{a}_j$	Potential energy between the electrons and nuclei. It is the sum of all electron-nucleus Coulomb interactions.
$\sum_{q>p=1}^{n} \dfrac{1}{r_{pq}}$	$\sum_{i,j,k,l} \langle i,j|\dfrac{1}{r_{ij}}|k,l\rangle\, \hat{a}_i^\dagger \hat{a}_j^\dagger \hat{a}_l \hat{a}_k$	Potential energy operator arising from electron-electron Coulomb repulsions.
$\sum_{B>A=1}^{M} \dfrac{Z_A Z_B}{r_{AB}}$	E_{NN}	Potential nuclear-nuclear (*NN*) repulsion energy operator; the sum of all nucleus-nucleus Coulomb repulsions.

Figure 4.4 – Molecular Hamiltonian as a linear combination of products of creation and annihilation operators

The weights of operators are given by the one-electron integrals using the HF method:

$$h_{i,j} = \langle i|\hat{h}|j\rangle = \int d\boldsymbol{r}_1 d\boldsymbol{s}_1\ \psi_i^*(\boldsymbol{r}_1)\chi_i^*(\boldsymbol{s}_1)\left(-\frac{1}{2}\nabla_{r_1}^2 - \sum_I \frac{Z_I}{|\boldsymbol{r}_1 - \boldsymbol{R}_I|}\right)\psi_j(\boldsymbol{r}_1)\chi_j(\boldsymbol{s}_1)$$

where $\boldsymbol{r}_i$ are the coordinates of electron i, $\boldsymbol{R}_I$ are the coordinates of atom I and Z_I is the atomic number of atom I, and the two-electron terms are given by:

$$g_{i,j,k,l} = \langle i,j|\hat{g}|k,l\rangle = \int d\boldsymbol{r}_1\ d\boldsymbol{r}_2 d\boldsymbol{s}_1\ d\boldsymbol{s}_2 \frac{\psi_i^*(\boldsymbol{r}_1)\chi_i^*(\boldsymbol{s}_1)\psi_j^*(\boldsymbol{r}_2)\chi_j^*(\boldsymbol{s}_2)\psi_k(\boldsymbol{r}_1)\chi_k(\boldsymbol{s}_1)\psi_l(\boldsymbol{r}_2)\chi_l(\boldsymbol{s}_2)}{|\boldsymbol{r}_1 - \boldsymbol{r}_2|}$$

The molecular Hamiltonian can be expressed in the basis of the solutions of the HF method $\{\psi_i(\boldsymbol{r})\}$, which are called **Molecular Orbitals (MOs)** [Panagiotis]:

We showed some example calculations of these integrals in *Section 2.3.6, Kinetic energy operation,* and *Section 2.3.7, Potential energy operation.* In the next section, we will see how to approximate the spatial wave functions in those integrals.

There are three commonly used HF methods:

- The **restricted HF (RHF)** method is used for closed-shell molecules. The spin-orbitals are either alpha (spin-up) or beta (spin-down) and all orbitals are doubly occupied by alpha and beta spin-orbitals.

- The restricted **open-shell (ROHF)** method is used for open-shell molecules where the numbers of electrons of each spin are not equal. ROHF uses as many doubly occupied molecular orbitals as possible and singly occupied orbitals for the unpaired electrons.

- The **unrestricted HF (UHF)** method is used for open-shell molecules where the numbers of electrons of each spin are not equal. UHF orbitals can have either alpha or beta spin, but the alpha and beta orbitals may have different spatial components.

4.5. Basis sets

The spatial wave functions, $\psi(r)$, in the integrals of *Section 4.4, Molecular Hamiltonian in the Hartree-Fock orbitals basis,* are approximated by linear combinations of several independent basis functions. The form of these functions is inspired by the atomic orbitals of hydrogen-like systems that we introduced in *Section 2.1.1, Spherical harmonic functions,* which have a radial part as shown in *Section 2.2.1, Computing the radial wave functions.*

Two classes of approximate basis orbitals that are commonly used are **Slater-type orbitals (STOs)** based on the Slater determinant introduced in *Section 4.2, Fock space*, and Cartesian **Gaussian-type orbitals (GTOs)**. These two types of basis functions can be combined as **STO-nG**, where n is the number of Gaussians used to make the approximation. **Ab Initio electronic structure** computations are conducted numerically using a basis set of orbitals.

We now detail the structure of these two classes and illustrate them with Python plots of functions.

4.5.1. Slater-type orbitals

STOs have the same structure as the atomic orbitals of hydrogen-like systems and their radial part have the following form [Wiki_GAU]:

$$R_l(r) = A(l, \alpha) r^l e^{-\alpha r}$$

where:

- l is the angular momentum quantum with values ranging from 0 to $n - 1$, where n

  is the principal quantum number.

- r is the nuclear distance of the electron from the atomic nucleus.

- α is called the orbital exponent and controls how fast the density of the orbital vanishes as a function of the nuclear distance.

$A(l, \alpha)$ is determined by the following normalization condition [Wiki-GAU]:

$$\int_0^\infty |R_l(r)|^2 \, r^2 dr = 1$$

$$A(l, \alpha)^2 \int_0^\infty (r^l e^{-\alpha r})^2 \, r^2 dr = 1$$

Noting that [Wiki-STO]:

$$\int_0^\infty x^l \, e^{-\alpha x} dx = \frac{l!}{\alpha^{n+1}}$$

we have:

$$A(l, \alpha) = (2\alpha)^{l+1} \sqrt{\frac{2\alpha}{(2l + 2)!}}$$

For the 1 s orbital, we have $l = 0$, hence $A(l, \alpha) = 2\alpha^{\frac{3}{2}}$, and the radial part of the Slater orbital is:

$$R(r) = 2\alpha^{\frac{3}{2}}e^{-\alpha r}$$

Let's plot this function with the following Python code:

```python
x = np.linspace(-5,5,num=1000)
r = abs(x)

alpha = 1.0

R = 2*alpha**(1.5)*np.exp(-alpha*r)

plt.figure(figsize=(4,3))
plt.plot(x,R,label="STO 1s H")
plt.legend()
```

Figure 4.5 shows the result:

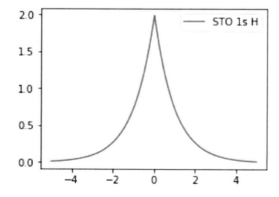

Figure 4.5 – Plot of the radial part of the Slater-type orbital for the 1s orbital of the hydrogen atom

We plot the antisymmetric spatial wave function for the hydrogen molecule as a linear combination of the preceding radial part of the slater orbital for a hydrogen atom as follows:

```
x = np.linspace(-7,7,num=1000)
r1 = abs(x+2.5)
r2 = abs(x-2.5)

alpha = 1.0

R = 2*alpha**(1.5)*np.exp(-alpha*r1)-2*alpha**(1.5)*np.exp(-alpha*r2)

plt.figure(figsize=(4,3))
plt.plot(x,R,label="Antisymmetric STO H2")
plt.legend()
```

Figure 4.6 shows the result:

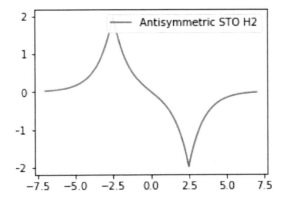

Figure 4.6 – Plot of the antisymmetric spatial wave function for the hydrogen molecule

4.5.2. Gaussian-type orbitals

GTOs have the same angular form as STOs, but their radial function adopts a Gaussian form [Wiki_GAU]:

$$R_l(r) = B(l,\alpha)r^l e^{-\alpha r^2}$$

where:

- l is the angular momentum quantum with values ranging from 0 to $n - 1$, where n is the principal quantum number.

- r is the nuclear distance of the electron from the atomic nucleus.

- α is called the orbital exponent and controls how fast the density of the orbital vanishes as a function of the nuclear distance.

$B(l, \alpha)$ is determined by the following normalization condition [Wiki-GAU]:

$$\int_0^\infty |R_l(r)|^2 r^2 dr = 1$$

In practice, we approximate the radial part of an STO with a linear combination of primitive Gaussian functions, called a contracted Gaussian function. The STO-nG basis sets include one contracted Gaussian function per atomic orbital [Skylaris]. We plot the STO-3G function for the $1s$ orbital of the hydrogen atom. Here is the code:

```
x = np.linspace(-7,7,num=1000)
r = abs(x)

c = [0.444635,0.535328,0.154329]
alpha = [0.109818,0.405771,2.227660]

psi = 0
for k in range(3):
    psi += c[k]*(2*alpha[k]/np.pi)**0.75 * np.exp(-alpha[k]*r**2)

plt.figure(figsize=(5,3))
plt.plot(x,psi,label="STO-3G 1s H")
plt.legend()
```

Figure 4.7 shows the result:

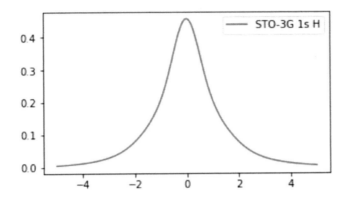

Figure 4.7 – Plot of the STO-3G function for the 1*s* orbital of the hydrogen atom

We plot the antisymmetric spatial wave function for the hydrogen molecule as a linear combination of the preceding radial part of the STO-3G function for the 1*s* orbital of a hydrogen atom as follows:

```python
x = np.linspace(-7,7,num=1000)
r1 = abs(x+2.5)
r2 = abs(x-2.5)

c = [0.444635,0.535328,0.154329]
alpha = [0.109818,0.405771,2.227660]

psi = 0
for k in range(3):
    psi += c[k]*(2*alpha[k]/np.pi)**0.75 * np.exp(-
alpha[k]*r1**2) \
    - c[k]*(2*alpha[k]/np.pi)**0.75 * np.exp(-alpha[k]*r2**2)

plt.figure(figsize=(5,3))
plt.plot(x,psi,label="Antisymmetric STO-3G H2")
plt.legend()
```

Figure 4.8 shows the result:

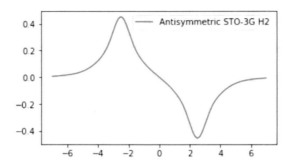

Figure 4.8 – Plot of the antisymmetric STO-3G function for the $1s$ orbital of the hydrogen molecule

4.6. Constructing a fermionic Hamiltonian with Qiskit Nature

The Qiskit Nature platform works with spin orbitals [Qiskit_Nat_1]. Each molecular orbital can have a spin-up or a spin-down electron, or spin-paired electrons. A spin orbital is either of those cases. For each molecular orbital, we have two spin orbitals. Let's now illustrate the construction of a fermionic Hamiltonian operator of the hydrogen molecule with Qiskit Nature.

4.6.1. Constructing a fermionic Hamiltonian operator of the hydrogen molecule

First, we define the molecular geometry of the hydrogen molecule with the Qiskit `Molecule` class, which has the following input parameters:

- `geometry`, a list of atom names, such as H for hydrogen, followed by Cartesian coordinates (x, y, z) of the atom's position in units of angstroms

- `charge`, an integer, the electric charge of the molecule

- `multiplicity`, an integer, the multiplicity $2S + 1$ of the molecule, where S is the total spin angular momentum, which is determined by the number of unpaired electrons in the molecule, that is, the number of electrons that occupy a molecular orbital singly, not with another electron:

```
hydrogen_molecule = Molecule(geometry=[['H', [0., 0., 0.]],
                             ['H', [0., 0., 0.735]]],
                    charge=0, multiplicity=1)
```

We define the electronic structure molecule driver by selecting the PySCF `driver` type and the basis set `sto3g`, which is the basis STO-3G we introduced in *Section 4.5.2, Gaussian-type orbitals*, in which the molecular orbitals are to be expanded. STO-3G is used by default in Qiskit Nature. RHF is used by default in Qiskit Nature's PySCF driver:

```
driver = ElectronicStructureMoleculeDriver(hydrogen_
molecule, basis='sto3g', driver_
type=ElectronicStructureDriverType.PYSCF)
```

We perform a HF calculation for the basis STO-3G. Here is the code:

```
qH2 = driver.run()
```

We create an `ElectronicStructureProblem` instance that produces the list of fermionic operators:

```
H2_fermionic_hamiltonian = ElectronicStructureProblem(driver)
```

We use the `second_q_ops()` method [Qiskit_Nat_3], which returns a list of second quantized operators: Hamiltonian operator, total particle number operator, total angular momentum operator, total magnetization operator, and, if available, x, y, z dipole operators:

```
H2_second_q_op = H2_fermionic_hamiltonian.second_q_ops()
```

Recall that in *Section 1.3, Quantum numbers and quantization of matter*, we introduced the **spin projection quantum number**, m_s, which gives the projection of the spin momentum s along the specified axis as either spin up (+½) or spin down (-½) in a given spatial direction. HF theory similarly defines α (up) and β (down) spin orbitals [Skylaris_1].

We define the `get_particle_number()` function, which gets the particle number property of a given electronic structure problem. Here is the code:

```
def get_particle_number(problem, show=True):
    particle_number = problem.grouped_property_transformed.get_
property("ParticleNumber")
    num_particles = (particle_number.num_alpha, particle_number.
num_beta)
    num_spin_orbitals = particle_number.num_spin_orbitals
    if show:
        print("Number of alpha electrons: {}".format(particle_
number.num_alpha))
```

```
    print("Number of beta electrons: {}".format(particle_
number.num_beta))
    print("Number of spin orbitals: {}".format(num_spin_
orbitals))
  return particle_number
```

We call the `get_particle_number()` function to get and print the particle number properties as follows:

```
print("Hydrogen molecule, basis: sto3g, Hartree-
Fock calculation")
H2_particle_number = get_particle_number(H2_fermionic_
hamiltonian)
```

Figure 4.9 shows the result where we see four spin orbitals, one α electron, and one β electron:

```
Hydrogen molecule, basis: sto3g, Hartree-Fock calculation
Number of alpha electrons: 1
Number of beta electrons: 1
Number of spin orbitals: 4
```

Figure 4.9 – Hydrogen molecule, HF calculation for the basis STO-3G, particle numbers

We define the `get_electronic_energy()` function, which returns the electronic energy property of a given electronic structure problem. Here is the code:

```
def get_electronic_energy(problem, show=True):
  electronic_energy = problem.grouped_property_transformed.get_
property("ElectronicEnergy")
  if show:
    print(electronic_energy)
  return electronic_energy
```

We call the `get_electronic_energy()` function to get and print the electronic energy as follows:

```
H2_electronic_energy = get_electronic_energy(H2_fermionic_
hamiltonian)
```

Figure 4.10 shows the molecular orbital (MO) one-body terms, where we see two α electron terms and two β electron terms:

```
(MO) 1-Body Terms:
        Alpha
        <(2, 2) matrix with 2 non-zero entries>
        [0, 0] = -1.2563390730032498
        [1, 1] = -0.47189600728114245
        Beta
        <(2, 2) matrix with 2 non-zero entries>
        [0, 0] = -1.2563390730032498
        [1, 1] = -0.47189600728114245
```

Figure 4.10 – Hydrogen molecule, electronic energy, molecular orbital (MO), one-body terms

Figure 4.11 shows the molecular orbital (MO) two-body terms that contain all possible spin combinations of molecular orbital (MO) two-body terms, $\alpha\alpha$, $\beta\alpha$, $\beta\beta$, $\alpha\beta$:

```
(MO) 2-Body Terms:
        Alpha-Alpha
        <(2, 2, 2, 2) matrix with 8 non-zero entries>
        [0, 0, 0, 0] = 0.6757101548035161
        [0, 0, 1, 1] = 0.6645817302552968
        [0, 1, 0, 1] = 0.18093119978423136
        [0, 1, 1, 0] = 0.18093119978423106
        [1, 0, 0, 1] = 0.18093119978423144
        ... skipping 3 entries
        Beta-Alpha
        <(2, 2, 2, 2) matrix with 8 non-zero entries>
        [0, 0, 0, 0] = 0.6757101548035161
        [0, 0, 1, 1] = 0.6645817302552968
        [0, 1, 0, 1] = 0.18093119978423136
        [0, 1, 1, 0] = 0.18093119978423106
        [1, 0, 0, 1] = 0.18093119978423144
        ... skipping 3 entries
        Beta-Beta
        <(2, 2, 2, 2) matrix with 8 non-zero entries>
        [0, 0, 0, 0] = 0.6757101548035161
        [0, 0, 1, 1] = 0.6645817302552968
        [0, 1, 0, 1] = 0.18093119978423136
        [0, 1, 1, 0] = 0.18093119978423106
        [1, 0, 0, 1] = 0.18093119978423144
        ... skipping 3 entries
        Alpha-Beta
        <(2, 2, 2, 2) matrix with 8 non-zero entries>
        [0, 0, 0, 0] = 0.6757101548035161
        [0, 0, 1, 1] = 0.6645817302552968
        [0, 1, 0, 1] = 0.18093119978423136
        [0, 1, 1, 0] = 0.18093119978423106
        [1, 0, 0, 1] = 0.18093119978423144
        ... skipping 3 entries
```

Figure 4.11 – Hydrogen molecule, electronic energy, molecular orbital (MO), two-body terms

The `FermionicOp` class [Qiskit_Nat_2] in the sparse label mode displays each term of the fermionic operator by a string of items separated by a space, starting with a label followed by an underscore, _, and by a positive integer representing the index of the fermionic mode. *Figure 4.12* shows the list of labels, the corresponding symbols, and the fermionic operator:

Label	Symbol	Fermionic operator
I	$\mathbb{1}$	Identity
-	$\hat{a}$	Annihilation
+	$\hat{a}^\dagger$	Creation
N	$\hat{a}^\dagger\hat{a}$	Particle occupation number
E	$\hat{a}\hat{a}^\dagger$	Emptiness (hole) position number

Figure 4.12 – List of labels used by the Qiskit FermionicOp class

The Qiskit `FermionicOp` class truncates the display of the fermionic Hamiltonian operator according to the maximum number of characters set by the `set_truncation()` method with a default value of 200 [Qiskit_Nat_T]. If the truncation value is set to 0, truncation is disabled. We set truncation to None with the `set_truncation(0)` method and then we print all 14 terms of the fermionic Hamiltonian operator of the hydrogen molecule:

```
# Set truncation to None
H2_second_q_op[0].set_truncation(0)
# Print the Fermionic operator
print("Hydrogen molecule")
print(H2_second_q_op[0])
```

Figure 4.13 shows the result:

```
Hydrogen molecule
Fermionic Operator
register length=4, number terms=14
  (0.18093119978423106+0j) * ( +_0 -_1 +_2 -_3 )
+ (-0.18093119978423128+0j) * ( +_0 -_1 -_2 +_3 )
+ (-0.18093119978423128+0j) * ( -_0 +_1 +_2 -_3 )
+ (0.18093119978423144+0j) * ( -_0 +_1 -_2 +_3 )
+ (-0.47189600728114245+0j) * ( +_3 -_3 )
+ (-1.2563390730032498+0j) * ( +_2 -_2 )
+ (0.4836505304710653+0j) * ( +_2 -_2 +_3 -_3 )
+ (-0.47189600728114245+0j) * ( +_1 -_1 )
+ (0.6985737227320175+0j) * ( +_1 -_1 +_3 -_3 )
+ (0.6645817302552965+0j) * ( +_1 -_1 +_2 -_2 )
+ (-1.2563390730032498+0j) * ( +_0 -_0 )
+ (0.6645817302552965+0j) * ( +_0 -_0 +_3 -_3 )
+ (0.6757101548035161+0j) * ( +_0 -_0 +_2 -_2 )
+ (0.4836505304710653+0j) * ( +_0 -_0 +_1 -_1 )
```

Figure 4.13 – Fermionic Hamiltonian operator of the hydrogen molecule

We now print, with the `FermionicOp.to_matrix` method, a matrix representation of the fermionic operator of the hydrogen molecule in the Fock basis where the basis states are ordered in increasing bitstring order as 0000, 0001, …, 1111. Here is the code:

```
print(H2_second_q_op[0].to_matrix())
```

Figure 4.14 shows the result:

```
(1, 1)      (-0.47189600728114245+0j)
(2, 2)      (-1.2563390730032498+0j)
(3, 3)      (-1.244584549813327+0j)
(4, 4)      (-0.47189600728114245+0j)
(5, 5)      (-0.24521829183026744+0j)
(10, 5)     (0.18093119978423106+0j)
(6, 6)      (-1.0636533500290957+0j)
(9, 6)      (0.18093119978423128-0j)
(7, 7)      (-0.35332510410715545+0j)
(8, 8)      (-1.2563390730032498+0j)
(6, 9)      (0.18093119978423128-0j)
(9, 9)      (-1.0636533500290957+0j)
(5, 10)     (0.18093119978423144+0j)
(10, 10)    (-1.8369679912029833+0j)
(11, 11)    (-1.1606317377577642+0j)
(12, 12)    (-1.244584549813327+0j)
(13, 13)    (-0.35332510410715534+0j)
(14, 14)    (-1.1606317377577637+0j)
(15, 15)    (0.2142782384194728+0j)
```

Figure 4.14 – Matrix representation of the fermionic Hamiltonian operator of the hydrogen molecule in the Fock basis

The fermionic Hamiltonian operator of the hydrogen molecule contains four particle number operators, which are shown in *Figure 4.15*:

$h_{i,j}$	Particle number operator $\hat{a}^\dagger \hat{a}$	+ 0	- 0	+ 1	- 1	+ 2	- 2	+ 3	- 3
-0.4719	$\hat{a}_3^\dagger \hat{a}_3$							X	X
-1.2563	$\hat{a}_2^\dagger \hat{a}_2$					X	X		
-0.4719	$\hat{a}_1^\dagger \hat{a}_1$			X	X				
-1.2563	$\hat{a}_0^\dagger \hat{a}_0$	X	X						

Figure 4.15 – Particle number operators of the fermionic Hamiltonian of the hydrogen molecule

The fermionic Hamiltonian operator of the hydrogen molecule contains ten two-electron exchange operators shown in *Figure 4.16*:

$g_{i,j,k,l}$	Exchange operators $\hat{a}_i^\dagger \hat{a}_j^\dagger \hat{a}_l \hat{a}_k$	+ 0	- 0	+ 1	- 1	+ 2	- 2	+ 3	- 3
0.1809	$\hat{a}_0^\dagger \hat{a}_2^\dagger \hat{a}_3 \hat{a}_1$	x			x	x			x
-0.1809	$\hat{a}_0^\dagger \hat{a}_3^\dagger \hat{a}_2 \hat{a}_1$	x			x		x	x	
-0.1809	$\hat{a}_1^\dagger \hat{a}_2^\dagger \hat{a}_3 \hat{a}_0$		x	x		x			x
0.1809	$\hat{a}_1^\dagger \hat{a}_3^\dagger \hat{a}_2 \hat{a}_0$		x	x			x	x	
0.4836	$\hat{a}_2^\dagger \hat{a}_3^\dagger \hat{a}_3 \hat{a}_2$					x	x	x	x
0.6986	$\hat{a}_1^\dagger \hat{a}_3^\dagger \hat{a}_3 \hat{a}_1$			x	x			x	x
0.6646	$\hat{a}_1^\dagger \hat{a}_2^\dagger \hat{a}_2 \hat{a}_1$			x	x	x	x		
0.6646	$\hat{a}_0^\dagger \hat{a}_3^\dagger \hat{a}_3 \hat{a}_0$	x	x					x	x
0.6757	$\hat{a}_0^\dagger \hat{a}_2^\dagger \hat{a}_2 \hat{a}_0$	x	x			x	x		
0.4836	$\hat{a}_0^\dagger \hat{a}_1^\dagger \hat{a}_1 \hat{a}_0$	x	x	x	x				

Figure 4.16 – Two-electron exchange operators of the fermionic Hamiltonian of the hydrogen molecule

Let's now illustrate the construction of a fermionic Hamiltonian operator of the lithium hydride molecule with Qiskit Nature.

4.6.2. Constructing a fermionic Hamiltonian operator of the lithium hydride molecule

We define the molecular geometry of the lithium hydride (LiH) molecule with the Qiskit Molecule class as we have explained in *Section 4.6.1, Constructing a fermionic Hamiltonian operator of the hydrogen molecule*:

```
LiH_molecule = Molecule(geometry=[['Li', [0., 0., 0.]],
                                  ['H', [0., 0., 1.5474]]],
                        charge=0, multiplicity=1)
```

We define the electronic structure molecule driver by selecting the PySCF `driver` type and the `sto3g` basis set in which the molecular orbitals are to be expanded:

```
driver = ElectronicStructureMoleculeDriver(LiH_
molecule, basis='sto3g', driver_
type=ElectronicStructureDriverType.PYSCF)
```

We create an `ElectronicStructureProblem` instance that produces the list of fermionic operators with the `freeze core=True` and `remove_orbitals=[4,3]` parameters, removing unoccupied orbitals:

```
LiH_fermionic_hamiltonian = ElectronicStructureProblem(driver,
  transformers=[FreezeCoreTransformer(freeze_core=True, remove_
orbitals=[4, 3])])
```

We use the `second_q_ops()` method to get a list of second quantized operators:

```
LiH_second_q_op = LiH_fermionic_hamiltonian.second_q_ops()
```

We call `get_particle_number()` to get and print the particle number property as follows:

```
print("Lithium hydride molecule, basis: sto3g, Hartree-
Fock calculation")
```

```
print("Parameters freeze_core=True, remove_orbitals=[4, 3]")
```

```
LiH_particle_number = get_particle_number(LiH_fermionic_
hamiltonian)
```

Figure 4.17 shows the result where we see six spin orbitals, one α electron, and one β electron:

```
Lithium hydride molecule, basis: sto3g, Hartree-Fock calculation
Parameters freeze_core=True, remove_orbitals=[4, 3]
Number of alpha electrons: 1
Number of beta electrons: 1
Number of spin orbitals: 6
```

Figure 4.17 – Lithium hydride molecule, HF calculation for the basis STO-3G, particle number

We call the `get_electronic_energy()` function to get and print the electronic energy as follows:

```
LiH_electronic_energy = get_electronic_energy(LiH_fermionic_
hamiltonian)
```

Figure 4.18 shows the molecular orbital (MO) one-body terms where we see two α electron terms and two β electron terms:

```
(MO) 1-Body Terms:
    Alpha
    <(3, 3) matrix with 9 non-zero entries>
    [0, 0] = -0.7806641144801698
    [0, 1] = 0.04770212338464205
    [0, 2] = -0.12958118897501983
    [1, 0] = 0.04770212338464208
    [1, 1] = -0.35909729348087005
    ... skipping 4 entries
    Beta
    <(3, 3) matrix with 9 non-zero entries>
    [0, 0] = -0.7806641144801698
    [0, 1] = 0.04770212338464205
    [0, 2] = -0.12958118897501983
    [1, 0] = 0.04770212338464208
    [1, 1] = -0.35909729348087005
    ... skipping 4 entries
```

Figure 4.18 – LiH molecule, electronic energy, molecular orbital (MO), one
-body terms

Figure 4.19 shows the molecular orbital (MO) two-body terms, which contains all possible spin combinations of molecular orbital two-body terms, $\alpha\alpha, \beta\alpha, \beta\beta, \alpha\beta$:

```
(MO) 2-Body Terms:
    Alpha-Alpha
    <(3, 3, 3, 3) matrix with 81 non-zero entries>
    [0, 0, 0, 0] = 0.4909697621018049
    [0, 0, 0, 1] = -0.04770211713792707
    [0, 0, 0, 2] = 0.1295811892794574
    [0, 0, 1, 0] = -0.04770211713792706
    [0, 0, 1, 1] = 0.2251551266886731
    ... skipping 76 entries
    Beta-Alpha
    <(3, 3, 3, 3) matrix with 81 non-zero entries>
    [0, 0, 0, 0] = 0.4909697621018049
    [0, 0, 0, 1] = -0.04770211713792707
    [0, 0, 0, 2] = 0.1295811892794574
    [0, 0, 1, 0] = -0.04770211713792706
    [0, 0, 1, 1] = 0.2251551266886731
    ... skipping 76 entries
    Beta-Beta
    <(3, 3, 3, 3) matrix with 81 non-zero entries>
    [0, 0, 0, 0] = 0.4909697621018049
    [0, 0, 0, 1] = -0.04770211713792707
    [0, 0, 0, 2] = 0.1295811892794574
    [0, 0, 1, 0] = -0.04770211713792706
    [0, 0, 1, 1] = 0.2251551266886731
    ... skipping 76 entries
    Alpha-Beta
    <(3, 3, 3, 3) matrix with 81 non-zero entries>
    [0, 0, 0, 0] = 0.4909697621018049
    [0, 0, 0, 1] = -0.04770211713792707
    [0, 0, 0, 2] = 0.1295811892794574
    [0, 0, 1, 0] = -0.04770211713792706
    [0, 0, 1, 1] = 0.2251551266886731
```

Figure 4.19 – LiH molecule, electronic energy, molecular orbital (MO), two-body terms

The Qiskit `FermionicOp` class truncates the display of the fermionic Hamiltonian operator according to the maximum number of characters set by the `set_truncation()` method with a default value of `200` [Qiskit_Nat_T]. If the truncation value is set to `0`, truncation is disabled. We set truncation to `1000` with the `set_truncation(1000)` method and then we print the first 20 terms of the more than a hundred terms of the fermionic operator of the LiH molecule:

```
# Set truncation to 1000
LiH_second_q_op[0].set_truncation(1000)
# Print the Fermionic operator
print("Lithium hydride molecule")
print(LiH_second_q_op[0])
```

Figure 4.20 shows the result:

```
Lithium hydride molecule
Fermionic Operator
register length=6, number terms=117
  (0.012557929502036855+0j) * ( +_0 -_1 +_3 -_4 )
+ (-0.0339966338792832+0j) * ( +_0 -_1 +_3 -_5 )
+ (-0.012557929502036855+0j) * ( +_0 -_1 -_3 +_4 )
+ (0.03399663387928322+0j) * ( +_0 -_1 -_3 +_5 )
+ (0.008886443232625232+0j) * ( +_0 -_1 +_4 -_5 )
+ (-0.008886443232625243+0j) * ( +_0 -_1 -_4 +_5 )
+ (0.047702123384813215+0j) * ( +_0 -_1 )
+ (-0.042727425131810944+0j) * ( +_0 -_1 +_5 -_5 )
+ (0.006789859849581841+0j) * ( +_0 -_1 +_4 -_4 )
+ (-0.0477021171380316+0j) * ( +_0 -_1 +_3 -_3 )
+ (-0.01129141206818323+0j) * ( +_0 -_1 +_2 -_2 )
+ (-0.0339966338792832+0j) * ( +_0 -_2 +_3 -_4 )
+ (0.12338438785306982+0j) * ( +_0 -_2 +_3 -_5 )
+ (0.033996633879283214+0j) * ( +_0 -_2 -_3 +_4 )
+ (-0.12338438785306986+0j) * ( +_0 -_2 -_3 +_5 )
+ (-0.03143601306362773+0j) * ( +_0 -_2 +_4 -_5 )
+ (0.03143601306362771+0j) * ( +_0 -_2 -_4 +_5 )
+ (-0.129581188974967+0j) * ( +_0 -_2 )
+ (0.1375589925618882+0j) * ( +_0 -_2 +_5 -_5 )
+ (-0.010949002449295066+0j) * ( +_0 -_2 +_4 -_4 )
+ (0.1295811892794 ...
```

Figure 4.20 – First 20 terms of the fermionic Hamiltonian operator of the lithium hydride molecule

We now print a matrix representation of the fermionic operator of the lithium hydride molecule in the Fock basis where the basis states are ordered in increasing bitstring order as 0000, 0001, …, 1111. Here is the code:

```
print(LiH_second_q_op[0].to_matrix())
```

Figure 4.21 shows the result:

```
(1, 1)        (-0.22617114899840962+0j)
(2, 1)        (0.06823802789335681+0j)
(4, 1)        (-0.12958118897501184+0j)
(1, 2)        (0.06823802789335663-0j)
(2, 2)        (-0.35909729348087616+0j)
(4, 2)        (0.04770212338466651+0j)
(3, 3)        (-0.36985826494689783+0j)
(5, 3)        (0.036410711316521155+0j)
(6, 3)        (0.149416634656926-0j)
(1, 4)        (-0.1295811889750119+0j)
(2, 4)        (0.04770212338466654-0j)
(4, 4)        (-0.7806641144801659+0j)
(3, 5)        (0.03641071131652117+0j)
(5, 5)        (-0.6744095760097841+0j)
(6, 5)        (0.05111732964052497+0j)
(3, 6)        (0.14941663465692606+0j)
(5, 6)        (0.05111732964052479+0j)
(6, 6)        (-0.9271642107743597+0j)
(7, 7)        (-0.60549949477159+0j)
(8, 8)        (-0.22617114899840962+0j)
(16, 8)       (0.06823802789335681+0j)
(32, 8)       (-0.12958118897501184+0j)
(9, 9)        (0.003031109061782944+0j)
(10, 9)       (0.024426933598162934+0j)
(12, 9)       (0.007977803586905058+0j)
 :     :
```

Figure 4.21 – Matrix representation of the fermionic Hamiltonian operator of the lithium hydride molecule in the Fock basis

4.7. Fermion to qubit mappings

We consider a system of N fermions, each labeled with an integer from 0 to $N - 1$. We need a fermion to qubit mapping, a description of the correspondence between states of fermions and states of qubits, or, equivalently, between fermionic operators and multi-qubit operators. We need a mapping between the fermion creation and annihilation operators and multi-qubit operators. The Jordan-Wigner and the Bravyi-Kitaev transformations are widely used and simulate a system of electrons with the same number of qubits as electrons.

4.7.1. Qubit creation and annihilation operators

We define qubit operators that act on local qubits [Yepez] [Chiew], as shown in *Figure 4.22*:

Qubit operator	Description
$\mathbb{1} = \begin{pmatrix} 1 & 0 \\ 0 & 1 \end{pmatrix}$	Identity
$\sigma^- = \begin{pmatrix} 0 & 1 \\ 0 & 0 \end{pmatrix} = \frac{1}{2}(\sigma_x + i\sigma_y) = \frac{1}{2}(X + iY) = \lvert 0 \rangle\langle 1 \rvert$	Annihilation
$\sigma^+ = \begin{pmatrix} 0 & 0 \\ 1 & 0 \end{pmatrix} = \frac{1}{2}(\sigma_x - i\sigma_y) = \frac{1}{2}(X - iY) = \lvert 1 \rangle\langle 0 \rvert$	Creation
$\sigma^+\sigma^- = \begin{pmatrix} 0 & 0 \\ 0 & 1 \end{pmatrix} = \frac{1}{2}(\mathbb{1} - \sigma_z) = \frac{1}{2}(\mathbb{1} - Z)$	One number (particle)
$\sigma^-\sigma^+ = \begin{pmatrix} 1 & 0 \\ 0 & 0 \end{pmatrix} = \frac{1}{2}(\mathbb{1} + \sigma_z) = \frac{1}{2}(\mathbb{1} + Z)$	Zero number (hole)

Figure 4.22 – Qubit creation and annihilation operators

The qubit operators have the anti-commutation relation: $\{\sigma^+, \sigma^-\} = \sigma^+\sigma^- + \sigma^-\sigma^+ = \mathbb{1}$.

4.7.2. Jordan-Wigner transformation

The **Jordan-Wigner** (**JW**) transformation stores the occupation of each spin orbital in each qubit. It maps the fermionic creation and annihilation operators to the tensor product of Pauli operators, as shown in *Figure 4.23* [Yepez] [Chiew] [Cao]. The operators σ_k^- and σ_k^+ change the occupation for the orbital level k. The tensor products of σ_z Pauli operators $\sigma_z^{\otimes k}$ enforce the fermionic anti-commutation relations by applying a phase according to the even or odd parity of the occupations for orbital labels less than k [Cao].

Operation	Fermionic operator	Qubit operator
Annihilation	$\hat{a}_0$	$\sigma^- = \frac{1}{2}(\sigma_x + i\sigma_y) = \frac{1}{2}(X + iY)$
	$\hat{a}_1$	$\frac{1}{2}(\sigma_x + i\sigma_y)\sigma_z = \frac{1}{2}(X + iY)Z = \frac{1}{2}XZ + \frac{i}{2}YZ$
	$\hat{a}_k$	$\mathbb{1}^{N-k-1} \otimes \sigma_k^+ \otimes \sigma_z^{\otimes k} = \frac{1}{2}\mathbb{1}^{N-k-1} \otimes (X + iY) \otimes Z^{\otimes k}$
Creation	$\hat{a}_0^\dagger$	$\sigma^+ = \frac{1}{2}(\sigma_x - i\sigma_y) = \frac{1}{2}(X - iY)$
	$\hat{a}_1^\dagger$	$\frac{1}{2}(\sigma_x - i\sigma_y)\sigma_z = \frac{1}{2}(X - iY)Z = \frac{1}{2}XZ - \frac{i}{2}YZ$
	$\hat{a}_k^\dagger$	$\mathbb{1}^{N-k-1} \otimes \sigma_k^+ \otimes \sigma_z^{\otimes k} = \frac{1}{2}\mathbb{1}^{N-k-1} \otimes (X - iY) \otimes Z^{\otimes k}$

Figure 4.23 – JW transformation

For example, for an orbital $k = 2$, we have the following mapping:

$$\hat{a}_2^\dagger \rightarrow \mathbb{1}^{N-3} \otimes \sigma^- \otimes \sigma_z^{\otimes 2} = \frac{1}{2}\mathbb{1}^{N-3} \otimes (X - iY) \otimes ZZ = \frac{1}{2}\mathbb{1}^{N-3}XZZ - \frac{i}{2}YZZ$$

The number of single Pauli operators σ_z scales linearly with the size of the system. The occupation number basis and the JW transformation allow the representation of a single fermionic creation or annihilation operator by $O(N)$ qubit operations.

The Hamiltonian that results from the JW transformation commutes with the number spin up and number spin down operators, which can be used to taper off two qubits [de Keijzer].

We define the `label_to_qubit()` function to convert a term of a fermionic operator represented as a sparse label to a qubit operator, which has the following input parameters:

- `label`, a sparse label as shown in *Figure 4.12*, used by the Qiskit `FermionicOp` class
- `converter`, either `JordanWignerMapper()`, `ParityMapper()` or `BravyiKitaevMapper()`

Here is the code:

```
def label_to_qubit(label, converter):
    qubit_converter = QubitConverter(converter)
    f_op = FermionicOp(label)
    qubit_op = qubit_converter.convert(f_op)
    return qubit_op
```

Now we convert the fermionic operators "+_0", "+_1", "+_2", "+_3", and "+_4" into qubit operators with the JW transformation with the following code:

```
for k in ("+_0", "+_1", "+_2", "+_3", "+_4"):
    qubit_op = label_to_qubit(k, JordanWignerMapper())
    print("{}:\n {}\n".format(k, qubit_op))
```

Figure 4.24 shows the result, which matches the expected outcome of the JW transformation, with the Qiskit tensor ordering of qubits:

```
+_0:            +_1:            +_2:             +_3:              +_4:
0.5 * X         0.5 * XZ        0.5 * XZZ        0.5 * XZZZ        0.5 * XZZZZ
+ -0.5j * Y     + -0.5j * YZ    + -0.5j * YZZ    + -0.5j * YZZZ    + -0.5j * YZZZZ
```

Figure 4.24 – JW transformation illustrated with "+_0", "+_1", "+_2", "+_3", and "+_4"

4.7.3. Parity transformation

The parity transformation is dual to the JW transformation: the parity operators are low-weight, while the occupation operators become high-weight [Bravyi][Cao]. *Figure 4.25* shows the mapping of the fermionic creation and annihilation operators to the tensor product of Pauli operators:

Operation	Fermionic operator	Qubit operator
Annihilation	$\hat{a}_0$	$\sigma^- = \frac{1}{2}(\sigma_x + i\sigma_y) = \frac{1}{2}(X + iY)$
	$\hat{a}_1$	$\frac{1}{2}(\sigma_x\sigma_z + i\sigma_y\mathbb{1}) = \frac{1}{2}(XZ + iY\mathbb{1}) = \frac{1}{2}XZ + \frac{i}{2}Y\mathbb{1}$
	$\hat{a}_k$	$\frac{1}{2}[\mathbb{1}^{k-2}\sigma_x{}^{\otimes N-k-1}\sigma_x\sigma_z + i\sigma_x{}^{\otimes N-k-1}\sigma_y\mathbb{1}^{k-1}] = \frac{1}{2}(\mathbb{1}^{k-2}X^{\otimes N-k-1}XZ + iX^{\otimes N-k-1}Y\mathbb{1}^{k-1})$
Creation	$\hat{a}_0^\dagger$	$\sigma^+ = \frac{1}{2}(\sigma_x - i\sigma_y) = \frac{1}{2}(X - iY)$
	$\hat{a}_1^\dagger$	$\frac{1}{2}(\sigma_x\sigma_z - i\sigma_y\mathbb{1}) = \frac{1}{2}(XZ - iY\mathbb{1}) = \frac{1}{2}XZ - \frac{i}{2}Y\mathbb{1}$
	$\hat{a}_k^\dagger$	$\frac{1}{2}[\mathbb{1}^{k-2}\sigma_x{}^{\otimes N-k-1}\sigma_x\sigma_z - i\sigma_x{}^{\otimes N-k-1}\sigma_y\mathbb{1}^{k-1}] = \frac{1}{2}(\mathbb{1}^{k-2}X^{\otimes N-k-1}XZ - iX^{\otimes N-k-1}Y\mathbb{1}^{k-1})$

Figure 4.25 – Parity transformation

Now we convert the fermionic operators "+_0", "+_1", "+_2", "+_3", and "+_4" into qubit operators with the parity transformation with the following code:

```
for k in ("+_0", "+_1", "+_2", "+_3", "+_4"):
    qubit_op = label_to_qubit(k, ParityMapper())
    print("{}:\n {}\n".format(k, qubit_op))
```

Figure 4.26 shows the result, which matches the expected outcome of the parity transformation, with the Qiskit tensor ordering of qubits:

```
+_0:            +_1:            +_2:            +_3:            +_4:
0.5 * X         -0.5j * YI      -0.5j * YII     -0.5j * YIII    -0.5j * YIIII
+ -0.5j * Y     + 0.5 * XZ      + 0.5 * XZI     + 0.5 * XZII    + 0.5 * XZIII
```

Figure 4.26 – Parity transformation illustrated with "+_0", "+_1", "+_2", "+_3", and "+_4"

The parity transformation introduces known symmetries that can be exploited to reduce the size of the problem by two qubits.

4.7.4. Bravyi-Kitaev transformation

The Bravyi-Kitaev (BK) transformation applies only for systems of N fermions where N is equal to a power of two, $N = 2^m$. The BK basis and transformation require only $O(log\ N)$ qubit operations to represent one fermionic operator. The BK transformation maps the occupation number basis $|f_0 \cdots f_{N-1}\rangle$ introduced in *Section 4.3, Fermionic creation and annihilation operators,* to the BK basis $|b_0 \cdots b_{N-1}\rangle$ with a matrix B_N, which is defined recursively [Cao][Seeley], where the sums are carried modulo 2:

$$B_1 = (1),\ B_{2^{m+1}} = \begin{pmatrix} B_{2^m} & \cdots & 11\ldots1 \\ \vdots & \ddots & \vdots \\ 0 & \cdots & B_{2^m} \end{pmatrix}$$

Hence:

$$B_2 = \begin{pmatrix} 1 & 1 \\ 0 & 1 \end{pmatrix},\ B_4 = \begin{pmatrix} 1 & 1 & 1 & 1 \\ 0 & 1 & 0 & 0 \\ 0 & 0 & 1 & 1 \\ 0 & 0 & 0 & 1 \end{pmatrix}$$

We define the `BK(m)` function, which returns a dictionary of matrices B_N for $N = 1$ to $N = 2^m$:

```
def BK(m):
  I = [[1, 0], [0, 1]]
  d = {}
  d[0] = [1]
  for k in range(0, m):
    B = np.kron(I,d[k])
    for l in range(2**k, 2**(k+1)):
      B[0,l] = 1
    d[k+1] = B
  return d
```

We compute the matrices B_1, B_2, B_4, and B_8 by calling the `BK(3)` function:

```
d = BK(3)
for k, v in d.items():
  s = "B_{"+str(2**k)+"} = "
  display(array_to_latex(v, prefix=s, precision = 0))
  print(" ")
```

Figure 4.27 shows the result:

$$B_1 = \begin{bmatrix} 1 \end{bmatrix}$$

$$B_2 = \begin{bmatrix} 1 & 1 \\ 0 & 1 \end{bmatrix}$$

$$B_4 = \begin{bmatrix} 1 & 1 & 1 & 1 \\ 0 & 1 & 0 & 0 \\ 0 & 0 & 1 & 1 \\ 0 & 0 & 0 & 1 \end{bmatrix}$$

$$B_8 = \begin{bmatrix} 1 & 1 & 1 & 1 & 1 & 1 & 1 & 1 \\ 0 & 1 & 0 & 0 & 0 & 0 & 0 & 0 \\ 0 & 0 & 1 & 1 & 0 & 0 & 0 & 0 \\ 0 & 0 & 0 & 1 & 0 & 0 & 0 & 0 \\ 0 & 0 & 0 & 0 & 1 & 1 & 1 & 1 \\ 0 & 0 & 0 & 0 & 0 & 1 & 0 & 0 \\ 0 & 0 & 0 & 0 & 0 & 0 & 1 & 1 \\ 0 & 0 & 0 & 0 & 0 & 0 & 0 & 1 \end{bmatrix}$$

Figure 4.27 – BK matrices B_1, B_2, B_4, and B_8

There are three sets to consider [Bravyi] [Mezzacapo] [Tranter]:

- The parity set $P(k)$ is the set of qubits that encodes the parity of the fermionic modes with an index less than k and that gives the global phase.

- The update set $U(k)$ is the set of qubits that must be flipped when the fermionic mode k changes occupation.

- The flip set $F(k)$ is the set of qubits that determines whether qubit k has the same or inverted parity with respect to the fermionic mode k. It is needed for odd k [Ribeiro].

These three sets can be obtained from the recursive matrices that map fermionic occupation to qubits. The remainder set $R(k) = P(k)\backslash F(k)$ is obtained from the set difference of the parity and flip sets [Bravyi].

Figure 4.28 shows the mapping of fermionic creation and annihilation operators to the tensor product of Pauli operators:

Operation	Fermionic operator	Qubit operator
Annihilation	$\hat{a}_0$	$\sigma^- = \frac{1}{2}(\sigma_x + i\sigma_y) = \frac{1}{2}(X + iY)$
	$\hat{a}_1$	$\frac{1}{2}(\sigma_x\sigma_z + i\sigma_y\mathbb{1}) = \frac{1}{2}(XZ + iY\mathbb{1}) = \frac{1}{2}XZ + \frac{i}{2}Y\mathbb{1}$
	$\hat{a}_k$	$\frac{1}{2}(X^{\otimes U(k)} \otimes X_k \otimes Z^{\otimes P(k)} + iX^{\otimes U(k)} \otimes Y_k \otimes Z^{\otimes R(k)})$
Creation	$\hat{a}_0^\dagger$	$\sigma^+ = \frac{1}{2}(\sigma_x - i\sigma_y) = \frac{1}{2}(X - iY)$
	$\hat{a}_1^\dagger$	$\frac{1}{2}(\sigma_x\sigma_z - i\sigma_y\mathbb{1}) = \frac{1}{2}(XZ - iY\mathbb{1}) = \frac{1}{2}XZ - \frac{i}{2}Y\mathbb{1}$
	$\hat{a}_k^\dagger$	$\frac{1}{2}(X^{\otimes U(k)} \otimes X_k \otimes Z^{\otimes P(k)} - iX^{\otimes U(k)} \otimes Y_k \otimes Z^{\otimes R(k)})$

Figure 4.28 – BK transformation

Now we convert the fermionic operators "+_0", "+_1", "+_2", "+_3", and "+_4" into qubit operators with the BK transformation with the following code:

```
for k in ("+_0", "+_1", "+_2", "+_3", "+_4"):
    qubit_op = label_to_qubit(k, BravyiKitaevMapper())
    print("{}:\n {}\n".format(k, qubit_op))
```

Figure 4.29 shows the result, which matches the expected outcome of the BK transformation, with the Qiskit tensor ordering of qubits:

```
+_0:              +_1:              +_2:              +_3:              +_4:
0.5 * X           -0.5j * YI        0.5 * XZI         -0.5j * YIII      0.5 * XZIII
+ -0.5j * Y       + 0.5 * XZ        + -0.5j * YZI     + 0.5 * XZZI      + -0.5j * YZIII
```

Figure 4.29 – BK transformation illustrated with "+_0", "+_1", "+_2", "+_3", and "+_4 "

4.8. Constructing a qubit Hamiltonian operator with Qiskit Nature

This section shows how to construct a qubit Hamiltonian operator with Qiskit Nature for the hydrogen molecule and the lithium hydride molecule.

We define the `fermion_to_qubit()` function to convert a fermionic operator to a qubit operator, which has the following input parameters:

- `f_op`, a fermionic operator obtained as explained in *Section 4.6, Constructing a fermionic Hamiltonian with Qiskit Nature*

- `mapper`, either `"Jordan-Wigner"` or `"Parity"` or `"Bravyi-Kitaev"`

- `truncate`, an integer to truncate the display of the Pauli list, which can be very large; set to `20` items by default

- `two_qubit_reduction`, Boolean, by default `False`, that determines whether to carry out two-qubit reduction when possible

- `z2symmetry_reduction`, by default `None`, that indicates whether a Z2 symmetry reduction should be applied to resulting qubit operators that are computed based on mathematical symmetries that can be detected in the operator [de Keijzer]

- `show`, set to `True` by default to display the name of the transformation and results

Here is the code:

```
def fermion_to_qubit(f_op, second_q_
op, mapper, truncate=20, two_qubit_reduction=False, z2symmetry_
reduction=None, show=True):
  if show:
    print("Qubit Hamiltonian operator")
  dmap = {"Jordan-
Wigner": JordanWignerMapper(), "Parity": ParityMapper(),
"Bravyi-Kitaev": BravyiKitaevMapper()}
  qubit_op = None
  qubit_converter = None
  for k, v in dmap.items():
    if k == mapper:
      if show:
```

```
      print("{} transformation ". format(mapper))
    qubit_converter = QubitConverter(v, two_qubit_
reduction=two_qubit_reduction, z2symmetry_reduction=z2symmetry_
reduction)
    if two_qubit_reduction:
        qubit_op = qubit_converter.convert(second_q_op[0], num_
particles=f_op.num_particles)
    else:
        qubit_op = qubit_converter.convert(second_q_op[0])
    n_items = len(qubit_op)
    if show:
        print("Number of items in the Pauli list:", n_items)
        if n_items <= truncate:
          print(qubit_op)
        else:
          print(qubit_op[0:truncate])
  return qubit_op, qubit_converter
```

We now show how to construct a qubit Hamiltonian operator of the hydrogen molecule.

4.8.1. Constructing a qubit Hamiltonian operator of the hydrogen molecule

First, we select the qubit mapper called `JordanWignerMapper()`:

```
print("Hydrogen molecule")
H2_qubit_op, qubit_converter = fermion_to_qubit(H2_fermionic_
hamiltonian, H2_second_q_op, "Jordan-Wigner", two_qubit_
reduction=True)
```

Figure 4.30 shows the result:

```
Hydrogen molecule
Qubit Hamiltonian operator
Jordan-Wigner tranformation
Number of items in the Pauli list: 15
-0.8105479805373281 * IIII
- 0.22575349222402358 * ZIII
+ 0.17218393261915543 * IZII
+ 0.12091263261776633 * ZZII
- 0.22575349222402358 * IIZI
+ 0.17464343068300436 * ZIZI
+ 0.16614543256382414 * IZZI
+ 0.17218393261915543 * IIIZ
+ 0.16614543256382414 * ZIIZ
+ 0.16892753870087904 * IZIZ
+ 0.12091263261776633 * IIZZ
+ 0.04523279994605781 * XXXX
+ 0.04523279994605781 * YYXX
+ 0.04523279994605781 * XXYY
+ 0.04523279994605781 * YYYY
```

Figure 4.30 – Qubit Hamiltonian operator of H2 with the JW transformation

Next, we use the qubit mapper called `ParityMapper()` with `two_qubit_reduction=True` to eliminate two qubits in the qubit Hamiltonian operator [Qiskit_Nat_4] [Qiskit_Nat_5]:

```
print("Hydrogen molecule")
H2_qubit_op, qubit_converter = fermion_to_qubit(H2_
fermionic_hamiltonian, H2_second_q_op, "Parity", two_qubit_
reduction=True)
```

Figure 4.31 shows the resulting qubit Hamiltonian operator works on two qubits. Recall that there are four spin orbitals, as shown in *Figure 4.9*, and a register length of four, as shown in *Figure 4.13* in *Section 4.6.1*, *Constructing a fermionic Hamiltonian operator of the hydrogen molecule*. A two-qubit reduction has been achieved:

```
Hydrogen molecule
Qubit Hamiltonian operator
Parity tranformation
Number of items in the Pauli list: 5
-1.0523732457728605 * II
+ (-0.39793742484317884+1.3877787807814457e-17j) * ZI
+ (0.39793742484317884-2.7755575615628914e-17j) * IZ
+ (-0.011280104256235116+1.3877787807814457e-17j) * ZZ
+ (0.18093119978423114-3.469446951953614e-18j) * XX
```

Figure 4.31 – Qubit Hamiltonian operator of H2 with parity transformation, two_qubit_reduction=True

Last, we select the qubit mapper called `BravyiKitaevMapper()`:

```
print("Hydrogen molecule")
H2_qubit_op, qubit_converter = fermion_to_qubit(H2_fermionic_
hamiltonian, H2_second_q_op, "Bravyi-Kitaev", two_qubit_
reduction=True)
```

Figure 4.32 shows the result:

```
Hydrogen molecule
Qubit Hamiltonian operator
Bravyi-Kitaev tranformation
Number of items in the Pauli list: 15
-0.8105479805373281 * IIII
+ 0.17218393261915543 * IZII
+ 0.12091263261776633 * IIZI
+ 0.12091263261776633 * ZIZI
- 0.22575349222402358 * ZZZI
+ 0.17218393261915543 * IIIZ
+ 0.16892753870087904 * IZIZ
+ 0.17464343068300436 * ZZIZ
- 0.22575349222402358 * IIZZ
+ 0.16614543256382414 * IZZZ
+ 0.16614543256382414 * ZZZZ
+ 0.04523279994605781 * IXIX
+ 0.04523279994605781 * ZXIX
- 0.04523279994605781 * IXZX
- 0.04523279994605781 * ZXZX
```

Figure 4.32 – Qubit Hamiltonian operator of H2 with the Bravyi-Kitaev transformation

4.8.2. Constructing a qubit Hamiltonian operator of the lithium hydride molecule

We use the qubit mapper called `ParityMapper()` with `two_qubit_reduction=True` to eliminate two qubits in the qubit Hamiltonian operator [Qiskit_Nat_4] [Qiskit_Nat_5]. We set `z2symmetry_reduction="auto"`. We print the first 20 items of the qubit Hamiltonian operator of the LiH molecule:

```
print("Lithium hydride molecule")
print("Using the ParityMapper with two_qubit_
reduction=True to eliminate two qubits")
print("Setting z2symmetry_reduction=\"auto\"")
LiH_qubit_op, qubit_converter = fermion_to_qubit(LiH_
fermionic_hamiltonian, LiH_second_q_op, "Parity", two_qubit_
reduction=True, z2symmetry_reduction="auto")
```

Figure 4.33 shows the resulting qubit Hamiltonian operator works on four qubits. Recall that there are six spin orbitals, as shown in *Figure 4.17*, and a register length of six, as shown in *Figure 4.20* in *Section 4.6.2, Constructing a fermionic Hamiltonian operator of the lithium hydride molecule*. A two-qubit reduction has been achieved:

```
Lithium hydride molecule
Using the ParityMapper with two_qubit_reduction=True to eliminate two qubits
Setting z2symmetry_reduction="auto"
Qubit Hamiltonian operator
Parity tranformation
Number of items in the Pauli list: 100
-0.20316606150558983 * IIII
+ (-0.365258690216036-2.7755575615628914e-17j) * ZIII
+ 0.09275994933497331 * IZII
- 0.2118898429700919 * ZZII
+ (0.3652586902160359-2.7755575615628914e-17j) * IIZI
- 0.11384335176465024 * ZIZI
+ 0.11395251883047082 * IZZI
+ (-0.06044012857315371-3.469446951953614e-18j) * ZZZI
+ (-0.0927599493349734+3.469446951953614e-18j) * IIIZ
+ 0.11395251883047082 * ZIIZ
- 0.12274244052544887 * IZIZ
+ 0.05628878167217043 * ZZIZ
+ (-0.2118898429700919-1.3877787807814457e-17j) * IIZZ
+ 0.06044012857315371 * ZIZZ
+ (-0.056288781672170426+3.469446951953614e-18j) * IZZZ
+ 0.08460131391824115 * ZZZZ
+ 0.019389408583701855 * XIII
+ (-0.01938940858370185-4.336808689942018e-19j) * XZII
+ (-0.0109527735737991-8.673617379884035e-19j) * XIZI
+ 0.010952773573799101 * XZZI
```

Figure 4.33 – Qubit Hamiltonian operator of LiH with parity transformation,
two_qubit_reduction=True

Summary

In this chapter, we have shown how to formulate an electronic structure program and map it into a qubit Hamiltonian, which is the input to a hybrid classical-quantum algorithm that is used to find the lowest energy eigenvalue for a quantum system. This is the topic of *Chapter 5, Variational Quantum Eigensolver (VQE)*.

Questions

Please test your understanding of the concepts presented in this chapter with the corresponding Google Colab notebook:

1. Which of the following terms is neglected in the BO approximation?

    A. Electronic kinetic energy operator.

    B. Nuclear kinetic energy operator.

    C. Potential energy between the electrons and nuclei. It is the sum of all electron-nucleus Coulomb interactions.

    D. Potential energy operator arising from electron-electron Coulomb repulsions.

    E. Potential nuclear-nuclear repulsion energy operator, the sum of all nucleus-nucleus Coulomb repulsions.

2. The Slater determinant wave function is antisymmetric with respect to:

    A. The exchange of two electrons (permutation of two rows)

    B. The exchange of two spin orbitals (permutation of two columns)

    C. Both of the above

3. Name three fermion to qubit transformations currently supported by Qiskit Nature.

4. Name two fermion to qubit transformations that simulate a system of electrons with the same number of qubits as electrons.

5. For which transformation does the resulting Hamiltonian commute with the number spin up and number spin down operators that can be used to taper off two qubits?

Answers

1. B

2. C

3. Jordan-Wigner, Parity, Bravyi-Kitaev

4. Jordan-Wigner, Parity

5. Jordan-Wigner

References

[Bravyi] Sergey Bravyi, Jay M. Gambetta, Antonio Mezzacapo, Kristan Temme, Tapering off qubits to simulate fermionic Hamiltonians, arXiv:1701.08213v1, January 27, 2017, `https://arxiv.org/pdf/1701.08213.pdf`

[Cao] Yudong Cao, Jonathan Romero, Jonathan P. Olson, Matthias Degroote, Peter D. Johnson, Mária Kieferová, Ian D. Kivlichan, Tim Menke, Borja Peropadre, Nicolas P. D. Sawaya, Sukin Sim, Libor Veis, Alán Aspuru-Guzik, Quantum Chemistry in the Age of Quantum Computing, Chem. Rev. 2019, 119, 19, 10856–10915, August 30, 2019, `https://doi.org/10.1021/acs.chemrev.8b00803`

[Chiew] Mitchell Chiew and Sergii Strelchuk, Optimal fermion-qubit mappings, arXiv:2110.12792v1 [quant-ph], October 25, 2021, `https://arxiv.org/pdf/2110.12792.pdf`

[De Keijzer] de Keijzer, R. J. P. T., Colussi, V. E., Škorić, B., and Kokkelmans, S. J. J. M. F. (2021), Optimization of the Variational Quantum Eigensolver for Quantum Chemistry Applications, arXiv, 2021, [2102.01781], `https://arxiv.org/abs/2102.01781`

[Grok] Grok the Bloch Sphere, `https://javafxpert.github.io/grok-bloch/`

[IBM_CEO] IBM CEO: Quantum computing will take off 'like a rocket ship' this decade, Fast Company, September 28, 2021, `https://www.fastcompany.com/90680174/ibm-ceo-quantum-computing-will-take-off-like-a-rocket-ship-this-decade`

[IBM_comp1] Welcome to IBM Quantum Composer, `https://quantum-computing.ibm.com/composer/docs/iqx/`

[IBM_comp2] IBM Quantum Composer, `https://quantum-computing.ibm.com/composer/files/new`

[Kaplan] Ilya G. Kaplan, Modern State of the Pauli Exclusion Principle and the Problems of Its Theoretical Foundation, Symmetry 2021, 13(1), 21, `https://doi.org/10.3390/sym13010021`

[Mezzacapo] Antonio Mezzacapo, Simulating Chemistry on a QuantumComputer, Part I, Qiskit Global Summer School 2020, IBM Quantum, Qiskit, Introduction to Quantum Computing and Quantum Hardware, https://qiskit.org/learn/intro-qc-qh/, Lecture Notes 8, https://github.com/qiskit-community/intro-to-quantum-computing-and-quantum-hardware/blob/master/lectures/introqcqh-lecture-notes-8.pdf?raw=true

[NumPy] NumPy: the absolute basics for beginners, https://numpy.org/doc/stable/user/absolute_beginners.html

[Panagiotis] Panagiotis Kl. Barkoutsos, Jerome F. Gonthier, Igor Sokolov, Nikolaj Moll, Gian Salis, Andreas Fuhrer, Marc Ganzhorn, Daniel J. Egger, Matthias Troyer, Antonio Mezzacapo, Stefan Filipp, Ivano Tavernelli, Quantum algorithms for electronic structure calculations: Particle-hole Hamiltonian and optimized wave-function expansions, Phys. Rev. A 98, 022322 – Published August 20, 2018, DOI: 10.1103/PhysRevA.98.022322, https://link.aps.org/doi/10.1103/PhysRevA.98.022322

[Qiskit] Qiskit, https://qiskit.org/

[Qiskit_Nat_0] Qiskit_Nature, https://github.com/Qiskit/qiskit-nature/blob/main/README.md

[Qiskit_Nat_1] Qiskit Nature and Finance Demo Session, with Max Rossmannek and Julien Gacon, October 15, 2021, https://www.youtube.com/watch?v=UtMVoGXlz04

[Qiskit_Nat_2] FermionicOp, https://qiskit.org/documentation/nature/stubs/qiskit_nature.operators.second_quantization.FermionicOp.html

[Qiskit_Nat_3] ElectronicStructureProblem.second_q_ops, https://qiskit.org/documentation/nature/stubs/qiskit_nature.problems.second_quantization.ElectronicStructureProblem.second_q_ops.html

[Qiskit_Nat_4] QubitConverter, https://qiskit.org/documentation/nature/stubs/qiskit_nature.converters.second_quantization.QubitConverter.html

[Qiskit_Nat_5] Qiskit Nature Tutorials, Electronic structure, https://qiskit.org/documentation/nature/tutorials/01_electronic_structure.html

[Qiskit_Nat_T] Second-Quantization Operators (qiskit_nature.operators.second_quantization) > FermionicOp > FermionicOp.set_truncation, https://qiskit.org/documentation/nature/stubs/qiskit_nature.operators.second_quantization.FermionicOp.set_truncation.html

[Qiskit_Nature] Introducing Qiskit Nature, Qiskit, Medium, April 6, 2021, `https://medium.com/qiskit/introducing-qiskit-nature-cb9e588bb004`

[Ribeiro] Sofia Leitão, Diogo Cruz, João Seixas, Yasser Omar, José Emilio Ribeiro, J.E.F.T. Ribeiro, Quantum Simulation of Fermionic Systems, CERN, `https://indico.cern.ch/event/772852/contributions/3505906/attachments/1905096/3146117/Quantum_Simulation_of_Fermion_Systems.pdf`

[Seeley] Jacob T. Seeley, Martin J. Richard, Peter J. Love, The Bravyi-Kitaev transformation for quantum computation of electronic structure, August 29, 2012, arXiv:1208.5986 [quant-ph], `https://arxiv.org/abs/1208.5986v1`

[Skylaris] CHEM6085: Density Functional Theory, Lecture 8, Gaussian basis sets, `https://www.southampton.ac.uk/assets/centresresearch/documents/compchem/DFT_L8.pdf`

[Skylaris_1] C.-K. Skylaris, CHEM3023: Spins, Atoms, and Molecules, Lecture 8, Experimental observables / Unpaired electrons, `https://www.southampton.ac.uk/assets/centresresearch/documents/compchem/chem3023_L8.pdf`

[Toulouse] Julien Toulouse, Introduction to quantum chemistry, January 20, 2021, `https://www.lct.jussieu.fr/pagesperso/toulouse/enseignement/introduction_qc.pdf`

[Tranter] Andrew Tranter, Peter J. Love, Florian Mintert, Peter V. Coveney, A comparison of the Bravyi-Kitaev and Jordan-Wigner transformations for the quantum simulation of quantum chemistry, December 5, 2018, J. Chem. Theory Comput. 2018, 14, 11, 5617–5630, `https://doi.org/10.1021/acs.jctc.8b00450`

[Wiki-Comb] Number of k-combinations for all k, Wikipedia, `https://en.wikipedia.org/wiki/Combination#Number_of_k-combinations_for_all_k`

[Wiki-GAU] Gaussian orbital, Wikipedia, `https://en.wikipedia.org/wiki/Gaussian_orbital`

[Wiki-STO] Slater-type orbital, Wikipedia, `https://en.wikipedia.org/wiki/Slater-type_orbital`

[Yepez] Jeffrey Yepez, Lecture notes: Quantum gates in matrix and ladder operator forms, January 15, 2013, `https://www.phys.hawaii.edu/~yepez/Spring2013/lectures/Lecture2_Quantum_Gates_Notes.pdf`

5

Variational Quantum Eigensolver (VQE) Algorithm

"Not only is the Universe stranger than we think, it is stranger than we can think."

– Werner Heisenberg

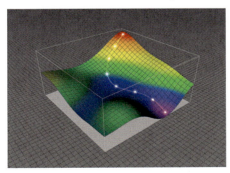

Figure 5.1 – Steepest descent line on a potential energy surface (PES) [authors]

We introduced the history behind the use of the variational method in *Section 1.1, Understanding the history of quantum chemistry and mechanics*. It is a mathematical construct that can be used computationally. Within the context of quantum chemistry, the variational method is used to determine the lowest energy associated with an eigenvalue, either the ground state or excited states.

The **Variational Quantum Eigensolver** (**VQE**) algorithm was introduced in 2014 [VQE_1] and is defined using quantum-based hardware. It is the first of several **Variational Quantum Algorithms** (**VQAs**) that are currently being explored by the scientific industry.

We use **Unitary Coupled Cluster Singles and Doubles** (**UCCSD**) as a starting point to determine a trial wave function for the variational method as it is essential that the VQE ansatz is close to the true ground state to make the VQE computations successful. To get an accurate energy estimate of 1 milli-Hartree (mHA), the ansatz for the VQE must be close to the true ground state by less than one in a million [Troyer]. In this chapter, we will focus on calculating only the ground state and Born-Oppenheimer potential energy surface (BOPES) for the hydrogen (H_2) and lithium hydride (LiH) molecules, and a macro molecule. We introduced the BOPES in *Section 4.1, Born-Oppenheimer approximation*. We will cover the following topics:

- *Section 5.1, Variational method*
- *Section 5.2, Example chemical calculations*

Technical requirements

A companion Jupyter notebook for this chapter can be downloaded from GitHub at `https://github.com/PacktPublishing/Quantum-Chemistry-and-Computing-for-the-Curious`, which has been tested in the Google Colab environment, which is free and runs entirely in the cloud, and in the IBM Quantum Lab environment. Please refer to *Appendix B – Leveraging Jupyter Notebooks in the Cloud*, for more information. The companion Jupyter notebook automatically installs the following list of libraries:

- **Numerical Python** (**NumPy**) [NumPy], an open-source Python library that is used in almost every field of science and engineering.

- Qiskit [Qiskit], an open-source SDK for working with quantum computers at the level of pulses, circuits, and application modules.

- Qiskit visualization support to enable the use of visualizations and Jupyter notebooks.

- Qiskit Nature [Qiskit_Nature] [Qiskit_Nat_0], a unique platform to bridge the gap between natural sciences and quantum simulations.

- **Python-based Simulations of Chemistry Framework (PySCF)** [PySCF], an open-source collection of electronic structure modules powered by Python.

- **Quantum Toolbox in Python (QuTiP)** [QuTiP], a general framework for solving quantum mechanics problems such as systems composed of few-level quantum systems and harmonic oscillators.

- **Atomic Simulation Environment (ASE)** [ASE_0], a set of tools and Python modules for setting up, manipulating, running, visualizing, and analyzing atomistic simulations. The code is freely available under the GNU LGPL license.

- PyQMC [PyQMC], a Python module that implements real-space quantum Monte Carlo techniques. It is primarily meant to interoperate with PySCF.

- h5py [h5py] package, a Pythonic interface to the HDF5 binary data format.

- SciPy [SciPy], a Python module that contains a large number of probability distributions, summary and frequency statistics, correlation functions and statistical tests, masked statistics, kernel density estimation, quasi-Monte Carlo functionality, and more.

Installing NumPy, Qiskit, QuTiP, and importing various modules

Install NumPy with the following command:

```
pip install numpy
```

Install Qiskit with the following command:

```
pip install qiskit
```

Install Qiskit visualization support with the following command:

```
pip install 'qiskit[visualization]'
```

Install Qiskit Nature with the following command:

```
pip install qiskit-nature
```

Install PySCF with the following command:

```
pip install pyscf
```

Install QuTiP with the following command:

```
pip install qutip
```

Install ASE with the following command:

```
pip install ase
```

Install PyQMC with the following command:

```
pip install pyqmc --upgrade
```

Install h5py with the following command:

```
pip install h5py
```

Install SciPy with the following command:

```
pip install scipy
```

Import NumPy with the following command:

```
import numpy as np
```

Import Matplotlib, a comprehensive library for creating static, animated, and interactive visualizations in Python, with the following command:

```
import matplotlib.pyplot as plt
```

Import the required functions and class methods. The `array_to_latex function()` returns a LaTeX representation of a complex array with dimension 1 or 2:

```
from qiskit.visualization import array_to_latex, plot_bloch_
vector, plot_bloch_multivector, plot_state_qsphere, plot_state_
city
from qiskit import QuantumRegister, ClassicalRegister,
QuantumCircuit, transpile
from qiskit import execute, Aer
import qiskit.quantum_info as qi
from qiskit.extensions import Initialize
from qiskit.providers.aer import extensions  # import aer
snapshot instructions
```

Import the Qiskit Nature libraries with the following commands:

```
from qiskit import Aer
from qiskit_nature.drivers import UnitsType, Molecule
from qiskit_nature.drivers.second_quantization
import ElectronicStructureDriverType,
ElectronicStructureMoleculeDriver
from qiskit_nature.problems.second_quantization import
ElectronicStructureProblem
from qiskit_nature.mappers.second_quantization import
ParityMapper, JordanWignerMapper, BravyiKitaevMapper
from qiskit_nature.converters.second_quantization import
QubitConverter
from qiskit_nature.transformers.second_quantization.electronic
import ActiveSpaceTransformer, FreezeCoreTransformer
from qiskit_nature.operators.second_quantization import
FermionicOp
from qiskit_nature.circuit.library.initial_states import
HartreeFock
from qiskit_nature.circuit.library.ansatzes import UCCSD
```

Import the Qiskit Nature property framework with the following command:

```
from qiskit_nature.properties import Property, GroupedProperty
```

Import the `ElectronicEnergy` property with the following command:

```
# https://qiskit.org/documentation/nature/tutorials/08_
property_framework.html
from qiskit_nature.properties.second_quantization.electronic
import (
    ElectronicEnergy,
    ElectronicDipoleMoment,
    ParticleNumber,
    AngularMomentum,
    Magnetization,
)
```

Import the `ElectronicIntegrals` property with the following command:

```
from qiskit_nature.properties.second_quantization.electronic.
integrals import (
    ElectronicIntegrals,
    OneBodyElectronicIntegrals,
    TwoBodyElectronicIntegrals,
    IntegralProperty,
)
from qiskit_nature.properties.second_quantization.electronic.
bases import ElectronicBasis
```

Import the Qiskit Aer state vector simulator and various algorithms with the following commands:

```
from qiskit.providers.aer import StatevectorSimulator
from qiskit import Aer
from qiskit.utils import QuantumInstance
from qiskit_nature.algorithms import VQEUCCFactory,
GroundStateEigensolver, NumPyMinimumEigensolverFactory,
BOPESSampler
from qiskit.algorithms import NumPyMinimumEigensolver, VQE,
HamiltonianPhaseEstimation
from qiskit.circuit.library import TwoLocal
from qiskit.algorithms.optimizers import QNSPSA
from qiskit.opflow import StateFn, PauliExpectation,
CircuitSampler, PauliTrotterEvolution
from functools import partial as apply_variation
```

Import the PySCF gto and scf libraries with the following command:

```
from pyscf import gto, scf
```

Import the PyQMC API library with the following command:

```
import pyqmc.api as pyq
```

Import h5py with the following command:

```
import h5py
```

Import the ASE libraries, the `Atoms` object, molecular data, and visualizations with the following commands:

```
from ase import Atoms
from ase.build import molecule
from ase.visualize import view
```

Import the math libraries with the following commands:

```
import cmath
import math
```

Import Python's statistical functions provided by the SciPy package with the following command:

```
import scipy.stats as stats
```

Import QuTiP with the following command:

```
import qutip
```

Import time and datetime with the following command:

```
import time, datetime
```

Import pandas and os.path with the following commands:

```
import pandas as pd
import os.path
```

5.1. Variational method

We illustrate the variational method through both classical and hybrid-quantum methods. We compare VQE to the variational Monte Carlo method. Further, we also compare the results for VQE to the **Quantum Phase Estimation** (**QPE**) algorithm, which is not a variational method.

In this section, we cover the following topics:

- *Section 5.1.1, The Rayleigh-Ritz variational theorem*
- *Section 5.1.2, Variational Monte Carlo methods*

- *Section 5.1.3, Quantum Phase Estimation (QPE)*
- *Section 5.1.4, Description of the VQE algorithm*

5.1.1. The Rayleigh-Ritz variational theorem

The Rayleigh-Ritz variational theorem states that the expectation value of the Hamiltonian $\widehat{H}$ of a system with respect to the state of an arbitrary wave function (Ψ) is always an upper bound to the exact ground state energy E_0 of the system it describes:

$$E_\Psi = \frac{\int \Psi^\dagger \widehat{H} \, \Psi d\tau}{\int \Psi^\dagger \, \Psi d\tau} \geq E_0$$

where τ generally represents time, spatial, and spin variables. This formula is not assuming any particular chemical setup nor reference frame.

We now give a proof of this theorem for the general Hamiltonian, which is represented by the discretized Hermitian operator $\widehat{H}$ [Toulouse]. Recall that, according to the spectral theorem introduced in *Section 2.3.1, Hermitian operator*, $\widehat{H}$ must have a set of orthonormal eigenvectors $\{|e_i\rangle ; i \in [0, K], \langle e_i|e_j\rangle = \delta_{ji}\}$ with real eigenvalues E_i, $\widehat{H} |e_i\rangle = E_i|e_i\rangle$ which form an orthonormal basis of the Hilbert space, and that $\widehat{H}$ has a unique spectral representation in this basis:

$$\widehat{H} = \sum_{i=0}^{K} E_i|e_i\rangle\langle e_i|$$

We can index the orthonormal eigenvectors of $\widehat{H}$ in increasing order of energy, $E_0 \leq E_1 \leq \cdots \leq E_K$ and decompose any state $|\Psi\rangle$ in this basis:

$$|\Psi\rangle = \sum_{k=0}^{K} c_k|e_k\rangle$$

with coefficients $c_k = \langle e_k|\Psi\rangle$ and the normalization constraint:

$$\langle\Psi|\Psi\rangle = \sum_{k=0}^{K} |c_k|^2 = 1$$

Noting that the complex conjugate transpose is:

$$\langle \Psi | = (|\Psi\rangle^*)^{\mathrm{T}} = \sum_{k=0}^{K} ((c_k|e_k\rangle)^*)^T = \sum_{k=0}^{K} c_k^* \langle e_k|$$

We compute the expectation value:

$$\langle \Psi|\hat{H}|\Psi\rangle = \sum_{k=0}^{K} \langle c_k^* e_k|\hat{H}_{elec}|c_k e_k\rangle = \sum_{k=0}^{K} c_k^* c_k E_k \langle e_k|e_k\rangle = \sum_{k=0}^{K} E_k |c_k|^2 \geq E_0$$

The minimum of this expression E_0 is reached for $c_0 = 1$ and $c_k = 0$ for all $k \geq 1$ that is for $|\Psi\rangle = |\Psi_0\rangle$. Hence:

$$E_\psi = \frac{\langle \Psi|\hat{H}|\Psi\rangle}{\langle \Psi|\Psi\rangle} \geq \frac{\langle \Psi_0|\hat{H}|\Psi_0\rangle}{\langle \Psi_0|\Psi_0\rangle} = E_0$$

To find the minimum, one can find the first and second derivative of the expression with respect to the parameters of the wave function. This setup for the variational theorem holds true for the electronic molecular Hamiltonian ($\hat{H}_{elec}$).

5.1.2. Variational Monte Carlo methods

The **variational Monte Carlo (VMC)** method is based on the Rayleigh-Ritz variational theorem [Chen] [Gorelov] [Toulouse_1] [Cao] [Dagrada] and Monte Carlo integration methods [Pease], noting that the expectation value can be rewritten in the form:

$$E_\psi = \langle \hat{H} \rangle_\psi = \frac{\int |\Psi|^2 [\Psi^{-1}\hat{H}\Psi] d\tau}{\int |\Psi|^2 d\tau} \geq E_0$$

We separate the integral into a probability distribution:

$$\mathcal{P}(\tau) = \frac{|\Psi|^2}{\int |\Psi|^2 d\tau}$$

and an observable:

$$\hat{E}(\tau) = \Psi^{-1}\hat{H}\Psi$$

which enables us to write the energy in the form of an average:

$$E_\Psi = \int \mathcal{P}(\boldsymbol{\tau})E(\boldsymbol{\tau})d\boldsymbol{\tau}$$

Now we apply an approximation to the E_Ψ formula, which is called the **Metropolis-Hastings (MH)** algorithm [Chen] [Toulouse1]. To perform the approximation mathematically, we sample a set of M points $\{(\boldsymbol{\tau})_k : k \in [1, M]\}$ from the probability distribution $\mathcal{P}(\boldsymbol{\tau})$ and we evaluate the local energy at each point $E(\boldsymbol{\tau})_k$, hence:

$$E_\Psi \approx \frac{1}{M}\sum_{k=1}^{M} E(\boldsymbol{\tau})_k$$

In practice, we can use a flexible explicitly correlated wave function Ψ.

We now give an illustration of the MH algorithm with Python code from Ref. [Stephens]. The MH algorithm is a **Markov chain Monte Carlo (MCMC)** method for producing samples from a probability distribution that we will call the target probability distribution. It works by simulating a Markov chain, whose stationary distribution is the target probability distribution. **Markov chain theory** is used to describe polymerization type reactions that are prominent in chemistry, chemical engineering, and in biology and medicine, such as the **polymerase chain reaction (PCR)** [Tamir].

We want to sample from the following probability distribution:

$$p(x) = \begin{cases} 0, & x < 0 \\ e^{-x}, & x \geq 0 \end{cases}$$

We implement the MH algorithm with a "random walk" kernel, $y = x + N(0,1)$, where N is the normal distribution, and the following acceptance probability:

$$A = min\left(1, \frac{p(y)}{p(x_t)}\right)$$

Here is the code:

```
n = 10000 # Size of the Markov chain stationary distribution

# Use np.linspace to create an array of n numbers between 0 and
n
index = np.linspace(0, n, num=n)
```

```
x = np.linspace(0, n, num=n)

x[0] = 3      # Initialize to 3
for i in range(1, n):
  current_x = x[i-1]

  # We add a N(0,1) random number to x
  proposed_x = current_x + stats.norm.rvs(loc=0, scale=1,
size=1, random_state=None)

  A = min(1, p(proposed_x)/p(current_x))

  r = np.random.uniform(0,1) # Generate a uniform random number
in [0, 1]

  if r < A:
    x[i] = proposed_x        # Accept move with probabilty
min(1,A)
  else:
    x[i] = current_x         # Otherwise "reject" move, and stay
where we are
```

We plot the locations visited by the Markov chain x:

```
plt.plot(index, x, label="Trace plot")
plt.xlabel('Index')
plt.ylabel('MH value')
plt.legend()
plt.show()
```

Figure 5.2 shows the result:

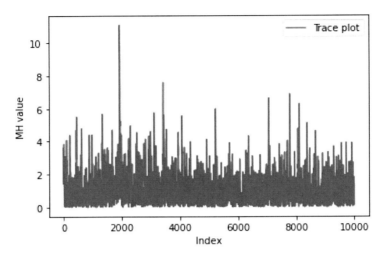

Figure 5.2 – Plot of the locations visited by the Markov chain x

We use the Freedman–Diaconis rule to select the "right" bin width to be used in a histogram [Bushmanov] [Freeman]:

```
q25, q75 = np.percentile(x, [25, 75])
bin_width = 2 * (q75 - q25) * len(x) ** (-1/3)
bins = round((x.max() - x.min()) / bin_width)
print("Freedman-Diaconis number of bins:", bins)
```

Here is the result:

```
Freedman-Diaconis number of bins: 109
```

We plot the histogram of the Markov chain x:

```
plt.hist(x, density=True, bins=bins)
plt.ylabel('Density')
plt.xlabel('x');
```

Figure 5.3 shows the result:

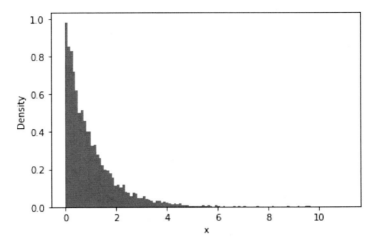

Figure 5.3 – Histogram of the Markov chain *x*

We see that the histogram of values of the Markov chain x is a good approximation to the distribution $p(x)$ defined previously.

We now define the `run_PySCF()` function, which computes the energy of the ground state with the PySCF RHF method and with the `OPTIMIZE` function in the PyQMC Python module that implements real-space variational Monte Carlo techniques [PyQMC]. It has the following parameters:

- `molecule`, the geometry of the molecule, defined with the Qiskit `Molecule` class
- `pyqmc`, set to `True` by default to run the PyQMC Python module
- `show`, set to `True` by default to display intermediate results
- Here is the definition of the `run_PySCF()` function:

```
def run_PySCF(molecule, pyqmc=True, show=True):
```

We now present the code that is contained in this `run_PySCF()` function. First, we reset the files:

```
# Reset the files
for fname in ['mf.hdf5','optimized_wf.hdf5']:
  if os.path.isfile(fname):
      os.remove(fname)
```

- Then we construct a PySCF molecular geometry from the molecule passed as an input parameter:

```
mol_PySCF = gto.M(atom = [" ".join(map(str, (name, *coord)))
for (name, coord) in molecule.geometry])
```

- We run the PySCF RHF method:

```
mf = scf.RHF(mol_PySCF)
mf.chkfile = "mf.hdf5"


conv, e, mo_e, mo, mo_occ = scf.hf.kernel(mf)
if show:
  if conv:
      print("PySCF restricted HF (RHF) converged ground-state
energy: {:.12f}".format(e))
  else:
      print("PySCF restricted HF (RHF) ground-state computation
failed to converge")
```

- Next, we run the `OPTIMIZE` function in the PyQMC Python module:

```
if pyqmc:
  pyq.OPTIMIZE("mf.hdf5",# Construct a Slater-Jastrow wave
function from the pyscf output
      "optimized_wf.hdf5", # Store optimized parameters in this
file.
      nconfig=100,            # Optimize using this many Monte
Carlo samples/configurations
      max_iterations=4,      # 4 optimization steps
      verbose=False)
```

- We read the content of the HDF5 file, which contains the optimized parameters, and if the PyQMC variational Monte Carlo computation converged, then we print the energy for each iteration:

```
with h5py.File("optimized_wf.hdf5") as f:
    iter = f['iteration']
    energy = f['energy']
    error = f['energy_error']
    l = energy.shape[0]
    e = energy[l-1]
    err = error[l-1]
    if show:
        if err < 0.1:
            print("Iteration, Energy, Error")
            for k in iter:
                print("{}:          {:.4f} {:.4f}".format(k,
energy[k], error[k]))
            print("PyQMC Monte Carlo converged ground-state
energy: {:.12f}, error: {:.4f}".format(e, err))
        else:
            print("PyQMC Monte Carlo failed to converge")
```

Finally, we let the run_PySCF() function return the following parameters to the caller:

- conv, Boolean, set to True if the PySCF RHF method converged
- e, the energy of the ground state
- Here is the return statement:

```
return conv, e
```

5.1.3. Quantum Phase Estimation (QPE)

In quantum chemistry, we need very accurate calculations of the total electronic energy of each molecule species involved in a chemical reaction [Burg]. The **Quantum Phase Estimation (QPE)** algorithm has a unique feature that it allows a bounded-error simulation of quantum systems, which makes it one of the most promising applications of future fault-tolerant quantum computing. Given a unitary operator U, its eigenstate and eigenvalues, $U|\psi\rangle = e^{2\pi i\theta}|\psi\rangle$, the ability to prepare a state $|\psi\rangle$, and the ability to apply U itself, the QPE algorithm calculates $2^n\theta$, where n is the number of qubits used to estimate θ thereby allowing measurement of θ as precisely as we want.

Recall that in *Section 2.5, Postulate 5 – Time evolution dynamics*, we saw that time evolution dynamics of a quantum system is described by Schrödinger's equation:

$$i\hbar\frac{d}{dt}|\psi\rangle = \hat{H}|\psi\rangle$$

For a time-independent Hamiltonian $\hat{H}$ with initial condition $|\psi(t_0)\rangle$, the solution is:

$$|\psi(t)\rangle = U(t)|\psi(t_0)\rangle$$

where $U(t) = exp\left(-i\frac{t}{\hbar}\hat{H}\right)$ is the unitary time-evolution operator. Further recall that any unitary matrix has eigenvalues of the form $e^{i\theta}$. An eigenvalue of $U(t)$ is also an eigenvalue of $\hat{H}$.

We now illustrate the use of the Qiskit `PhaseEstimation` class. First, we define a function $U(\theta)$, which creates a quantum circuit with a single qubit $|q_0\rangle$ and applies the following unitary:

$$U(\theta)|q_0\rangle = e^{2\pi i\theta}|q_0\rangle = p(2\pi\theta)|q_0\rangle$$

where $p(\lambda)$ is the gate we introduced in *Section 3.2.1, Single qubit quantum gates*, which has the matrix form:

$$p(\lambda) = \begin{pmatrix} 1 & 0 \\ 0 & e^{i\lambda} \end{pmatrix}$$

Here is the code:

```
def U(theta):
    unitary = QuantumCircuit(1)
    unitary.p(np.pi*2*theta, 0)
    return unitary
```

We define the do_qpe() function, which illustrates the use of the Qiskit Nature PhaseEstimation class, and which has three parameters:

- unitary, a function that implements a unitary
- nqubits, the number of qubits, by default 3
- show, set to True by default to display the phase returned by PhaseEstimation class

Here is the code:

```
def do_qpe(unitary, nqubits=3, show=True):
    state_in = QuantumCircuit(1)
    state_in.x(0)
    pe = PhaseEstimation(num_evaluation_qubits=nqubits, quantum_
instance=quantum_instance)
    result = pe.estimate(unitary, state_in)
    phase_out = result.phase
    if show:
        print("Number of qubits: {}, QPE phase estimate: {}".
format(nqubits, phase_out))
    return(phase_out)
```

First, we run a test of accuracy with three qubits:

```
quantum_instance = QuantumInstance(backend = Aer.get_
backend('aer_simulator_statevector'))
theta = 1/2 + 1/4 + 1/8
print("theta: {}".format(theta))
unitary = U(theta)
result = do_qpe(unitary, nqubits=3)
```

Here is the result:

```
theta: 0.875
Number of qubits: 3, QPE phase estimate: 0.875
```

Next, we run a test of accuracy with eight qubits:

```
theta = 1/2 + 1/4 + 1/8 + 1/16 + 1/32 + 1/64 + 1/128 + 1/256
print("theta: {}".format(theta))
```

```
unitary = U(theta)
result = do_qpe(unitary, nqubits=8)
```

Here is the result:

```
theta: 0.99609375
Number of qubits: 8, QPE phase estimate: 0.99609375
```

We see that we can get an estimate of the phase with a bounded error from the true phase by increasing the number of qubits that the `PhaseEstimation` class is allowed to use.

5.1.4. Description of the VQE algorithm

In a loop, a classical computer optimizes the parameters of a quantum circuit with respect to an objective function, such as finding the ground state of a molecule, which is the state with the lowest energy. The parameterized quantum circuit prepares a trial quantum state as a trial solution (an ansatz). By repeatedly measuring qubits at the output of the quantum circuit, we get the expectation value of the energy observable with respect to the trial state.

The VQE algorithm provides an estimate of the ground state of a given quantum system encoded as a Hamiltonian $\hat{H}$, the state of the system with the lowest energy E_0, for instance, the ground state energy of a molecule. It involves an iterative minimization of the expectation value $E_{\Psi(\theta)}$ of the energy observable with respect to the parametrized (θ) trial state $|\Psi(\theta)\rangle$:

$$E_0 \leq E_{\Psi(\theta)} = \langle \Psi(\theta)|\hat{H}|\Psi(\theta)\rangle$$

As shown in *Section 3.1.6, Pauli matrices*, we can decompose the Hamiltonian $\hat{H}$ into the weighted sum of M tensor products $P_k = \otimes_j^N \sigma_{i,j}$, where $\sigma_{i,j} \in \{\mathbb{1}, \sigma_x, \sigma_y, \sigma_z\}$ with weights c_k and N qubits:

$$\hat{H} = \sum_{k=0}^{M-1} c_k P_k = \sum_{k=0}^{M-1} c_k \otimes_j^N \sigma_{k,j}$$

Hence the expectation value of the energy observable $E_{\psi(\theta)}$ can be rewritten as follows:

$$E_{\Psi(\theta)} = \langle \Psi(\theta)|\hat{H}|\Psi(\theta)\rangle = \sum_{k=0}^{M-1} c_k \langle \Psi(\theta)|P_k|\Psi(\theta)\rangle = \sum_{k=0}^{M-1} c_k \langle \Psi(\theta)| \otimes_j^N \sigma_{k,j}|\Psi(\theta)\rangle$$

We prepare a trial state $|\Psi(\theta)\rangle$ with the set of parameters $\theta = (\theta_0, \theta_1, \dots, \theta_m)$ with a quantum circuit initialized in the state $|0\rangle^{\otimes N}$, and represented by $U(\theta)$, which outputs the state $|\Psi(\theta)\rangle = U(\theta)|0\rangle^{\otimes N}$.

By transposing the complex conjugate, $\langle\Psi| = \langle 0|^{\otimes N} U(\theta)^\dagger$, we can rewrite the expectation value of the energy observable $E_{\Psi(\theta)}$ as follows:

$$E_{\Psi(\theta)} = \langle\Psi(\theta)|\hat{H}|\Psi(\theta)\rangle = \langle 0|^{\otimes N} U(\theta)^\dagger \hat{H}\, U(\theta)|0\rangle^{\otimes N}$$

and then by taking the sum out to the front:

$$E_{\Psi(\theta)} = \sum_{k=0}^{M-1} \langle\Psi(\theta)|P_k|\Psi(\theta)\rangle = \sum_{k=0}^{M-1} c_k \langle 0|^{\otimes N} U(\theta)^\dagger \otimes_j^N \sigma_{k,j} U(\theta)|0\rangle^{\otimes N}$$

For each P_k, we run the quantum circuit $U(\theta)$ followed by rotations $R_k \in \{\mathbb{1}, R_X(-\pi/2), R_Y(\pi/2)\}$ depending on P_k before measuring the qubits in the Z basis so that we effectively measure the output state in the basis of the eigenvectors of P_k to get the expectation value $\langle\Psi(\theta)|P_k|\Psi(\theta)\rangle$ with respect to the output state $|\Psi(\theta)\rangle = U(\theta)|0\rangle^{\otimes N}$.

On a classical computer, we compute the weighted sum of the expectation values $\langle\Psi(\theta)|P_k|\Psi(\theta)\rangle$ with weights c_k to get the expectation value $E_{\Psi(\theta)}$ with respect to the output state $|\Psi(\theta)\rangle$. We update the set of parameters θ using a classical optimization routine, minimizing the expectation value $E_{\Psi(\theta)}$ until convergence in the value of the energy or the maximum allowable number of iterations is reached. The parameters θ_{min} at convergence define approximately the ground state $|\Psi(\theta_{min})\rangle$ of the quantum system encoded into a Hamiltonian $\hat{H}$ with the lowest energy $E_{\Psi(\theta_{min})} = \langle\Psi(\theta_{min})|\hat{H}|\Psi(\theta_{min})\rangle$. The algorithm is summarized in *Figure 5.4*.

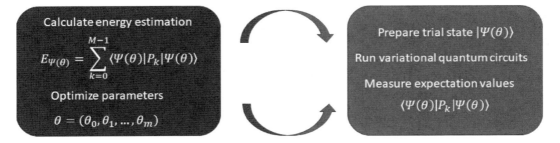

Figure 5.4 – VQE algorithm

Trial wave functions

The **Coupled-Cluster** (CC) theory constructs a multi-electron wave function (Ψ) using the exponential cluster operator $\hat{T} = \hat{T}_1 + \hat{T}_2 + \cdots \hat{T}_n$, where $\hat{T}_1$ is the operator for all single excitations, $\hat{T}_2$ is the operator for all double excitations, and so on. We start the VQE with the following **unitary Coupled-Cluster** (UCC) ansatz of the quantum state $|\Psi(\theta)\rangle$ with variational parameter θ [Panagiotis] [Lolur]:

$$|\Psi(\theta)\rangle = e^{\hat{T}(\theta) - \hat{T}^\dagger(\theta)} |\Psi_{ref}\rangle$$

where $|\Psi_{ref}\rangle$ is the Hartree-Fock ground state. In the UCC method restricted to the extension to single and double excitations (UCCSD), the operators $\hat{T}_1$ and $\hat{T}_2$ can be expanded as follows:

$$\hat{T}_1(\theta) = \sum_{i;m} \theta_i^m a_m^\dagger a_i$$

$$\hat{T}_2(\theta) = \frac{1}{2} \sum_{i,j;m,n} \theta_{i,j}^{m,n} a_n^\dagger a_m^\dagger a_j a_i$$

where:

- $\hat{a}_m^\dagger$ is the fermionic creation operator introduced in *Section 4.3.1, Fermion creation operator*
- $\hat{a}_i$ is the fermionic annihilation operator introduced in Section *4.3.2, Fermion annihilation operator*
- θ is the set of parameters for all expansion coefficients.

The UCCSD ansatz of the quantum state $|\Psi(\theta)\rangle$ is then mapped to qubit operators with the Jordan-Wigner (JW), the Parity, or the Bravyi-Kitaev (BK) transformation introduced in *Section 4.7, Fermion to qubit mappings,* resulting in an initial qubit state vector for the VQE calculation.

Setting up the VQE solver

We set up a noise-free simulation with the Qiskit Aer state vector simulator backend:

```
quantum_instance = QuantumInstance(backend = Aer.get_
backend('aer_simulator_statevector'))
```

Now we set up solving methods. To learn how to train circuit-based variational models, check Ref. [Qiskit_2021_Lab4]. First, we set up the NumPy minimum eigensolver as follows:

```
numpy_solver = NumPyMinimumEigensolver()
```

We set up the Two-Local circuit [Panagiotis] as follows:

```
tl_circuit = TwoLocal(rotation_blocks = ['h', 'rx'],
entanglement_blocks = 'cz',
                       entanglement='full', reps=2, parameter_
prefix = 'y')
```

We set up the VQE using a heuristic ansatz, the Two-Local circuit with the default **Sequential Least Squares Programming (SLSQP)** optimizer:

```
vqe_tl_solver = VQE(ansatz = tl_circuit,
                    quantum_instance = QuantumInstance(Aer.
get_backend('aer_simulator_statevector')))
```

Then we set up a solver with the **Unitary Coupled Cluster (UCC)** factory. It allows a fast initialization of a VQE initializing the qubits in the Hartree-Fock state and using the quantum UCC with singles and doubles (q-UCCSD), a popular wave function ansatz [VQE_2] [VQE_3]. Here is the code:

```
vqe_ucc_solver = VQEUCCFactory(quantum_instance, ansatz=tl_
circuit)
```

We set up a callback function, qnspsa_callback(), for the **Quantum Natural SPSA (QN-SPSA)** optimizer, which appends results to the loss array qnspsa_loss:

```
qnspsa_loss = []
def qnspsa_callback(nfev, x, fx, stepsize, accepted):
    qnspsa_loss.append(fx)
```

Now we are ready to show some examples.

5.2. Example chemical calculations

As discussed in *Chapter 4, Molecular Hamiltonians*, approximating the PES of nuclear motion occurs due to the use of the BO approximation. We can use a semi-empirical method of approximating the PES through experimental data and/or computer simulations.

The PES can be compared to a landscape with mountains and valleys. In practice, as chemists, we want to find the global minimum (ocean floor) not local minima (mountain meadows) of the PES, as seen in *Figure 5.1*. We use the variational method, both classical and quantum, to find the global minimum. This can be compared to a ball rolling around the landscape. If we give the ball a nudge in some direction, generally downward, the ball will wind up in the minimum. We call this gradient descent. The gradient descent can be supplied by numerically changing input values or by an analytic formula of the wave function that describes the PES.

To state that calculation of determining the PES we guess a trial wave function, which can be optimized in the calculation to enable us to find the global minimum of the energy. We call this global minimum the lowest energy possible for a given eigenvalue.

We present several implementations of solving for the ground state and plotting the BOPES of three molecules with the classical PySCF RHF, PyQMC variational Monte Carlo, the QPE, and the VQE with Qiskit Nature using the STO-3G basis with the PySCF driver.

In this section, we cover the following topics:

- *Section 5.2.1, Hydrogen molecule*

- *Section 5.2.2, Lithium hydride molecule*

- *Section 5.2.3, Macro molecule*

We use the `get_particle_number()` function defined in *Section 4.6.1, Constructing a fermionic Hamiltonian operator of the hydrogen molecule*, which gets the particle number property of a given electronic structure problem.

We use the `fermion_to_qubit()` function defined in *Section 4.8, Constructing a qubit Hamiltonian with Qiskit Nature,* to convert a fermionic operator to a qubit operator. It has the following input parameters:

- `f_op`, a fermionic operator obtained as explained in *Section 4.6, Constructing a fermionic Hamiltonian with Qiskit Nature*

- `mapper`, either "Jordan-Wigner" or "Parity" or "Bravyi-Kitaev"

- `truncate`, an integer to truncate Pauli list by default set to 20 items

- `two_qubit_reduction`, a Boolean, by default `False`, that determines whether to carry out two-qubit reduction when possible

- `z2symmetry_reduction`, by default `None`, that indicates whether a Z2 symmetry reduction should be applied to resulting qubit operators that are computed based on mathematical symmetries that can be detected in the operator [de Keijzer]

- `show`, set to `True` by default to display the name of the transformation and results

Qiskit Nature provides a class called `GroundStateEigensolver` to calculate the ground state of a molecule. We define the `run_vqe()` function, which has the following input parameters:

- `name`, a string of characters to be printed, such as `'NumPy exact solver'`

- `f_op`, a fermionic operator obtained as explained in *Section 4, Constructing a fermionic Hamiltonian with Qiskit Nature*

- `qubit_converter`, either `JordanWignerMapper()`, `ParityMapper()`, or `BravyiKitaevMapper()`, which is the output of the `fermion_to_qubit()` function

- `solver`, either one of the solvers defined in *Section 5.2.3, Setting up the VQE solver*, `numpy_solver`, `vqe_ucc_solver`, or `vqe_tl_solver`.

Here is the code:

```
def run_vqe(name, f_op, qubit_converter, solver, show=True):
  calc = GroundStateEigensolver(qubit_converter, solver)
  start = time.time()
  ground_state = calc.solve(f_op)
  elapsed = str(datetime.timedelta(seconds = time.time()-
start))
  if show:
    print("Running the VQE using the {}".format(name))
    print("Elapsed time: {} \n".format(elapsed))
    print(ground_state)
  return ground_state
```

We define the `run_qpe()` function to perform a quantum phase estimation and return an eigenvalue of a Hamiltonian as an estimation of the electronic ground state energy. It has the following input parameters:

- `particle_number`, the property returned by the `get_particle_number()` function

- `qubit_converter`, either `JordanWignerMapper()`, `ParityMapper()`, or `BravyiKitaevMapper()`, which is the output of the `fermion_to_qubit()` function

- `qubit_op`, a qubit Hamiltonian operator returned the `fermion_to_qubit()` function

- `n_ancillae`, an integer that defaults to 3, which is the number of ancillae qubits

- `num_time_slices`, an integer that defaults to 1, which is the number of **Trotterization** repetitions to make to improve the approximation accuracy, and is used by the Qiskit `PauliTrotterEvolution` class

- `show`, set to `True` by default to display intermediate results

Here is the code:

```
def run_qpe(particle_number, qubit_converter, qubit_op, n_
ancillae=3, num_time_slices = 1, show=True):
   initial_state = HartreeFock(particle_number.num_spin_
orbitals,
                             (particle_number.num_alpha,
                             particle_number.num_beta), qubit_
converter)

   state_preparation = StateFn(initial_state)

   evolution = PauliTrotterEvolution('trotter', reps=num_time_
slices)

   qpe = HamiltonianPhaseEstimation(n_ancillae, quantum_
instance=quantum_instance)
   result = qpe.estimate(qubit_op, state_preparation,
evolution=evolution)
```

```
    if show:
        print("\nQPE initial Hartree Fock state")
        display(initial_state.draw(output='mpl'))
        eigv = result.most_likely_eigenvalue
        print("QPE computed electronic ground state energy
(Hartree): {}".format(eigv))

    return eigv
```

We define the `plot_energy_landscape()` function to plot the energy as a function of atomic separation:

```
def plot_energy_landscape(energy_surface_result):
    if len(energy_surface_result.points) > 1:
        plt.plot(energy_surface_result.points, energy_surface_
result.energies, label="VQE Energy")
        plt.xlabel('Atomic distance Deviation(Angstrom)')
        plt.ylabel('Energy (hartree)')
        plt.legend()
        plt.show()
    else:
        print("Total Energy is: ", energy_surface_result.
energies[0], "hartree")
        print("(No need to plot, only one configuration
calculated.)")
    return
```

We define the `plot_loss()` function, which accepts the following input parameters:

- `loss`, an array of floats, optional, generated by the callback function
- `label`, a character string to be displayed by the `plot_loss()` function
- `target`, a float to be displayed by the `plot_loss()` function

Here is the code:

```
def plot_loss(loss, label, target):
    plt.figure(figsize=(12, 6))
    plt.plot(loss, 'tab:green', ls='--', label=label)
    plt.axhline(target, c='tab:red', ls='--', label='target')
```

```
plt.ylabel('loss')
plt.xlabel('iterations')
plt.legend()
```

We now define the `solve_ground_state()` function, which solves for a ground state. It accepts as input the following parameters, which define the geometry of the molecule:

- `molecule`, the geometry of the molecule, and the output of the `Molecule` function.

- `mapper`, either "`Jordan-Wigner`" or "`Parity`" or "`Bravyi-Kitaev`".

- `num_electrons`, an integer, optional, number of electrons for the `ActiveSpaceTransformer`. Defaults to 2.

- `num_molecular_orbitals`, an integer, optional, number of electron orbitals for `ActiveSpaceTransformer`. Defaults to 2.

The following list of input parameters control the whole process:

- `transformers`, an optional list of transformers. For example, for lithium hydride, we will use the following: `transformers=[FreezeCoreTransformer(freeze_core=True, remove_orbitals=[4, 3])]`.

- `two_qubit_reduction`, a Boolean, by default `False`. It determines whether to carry out two-qubit reduction when possible.

- `z2symmetry_reduction`, by default `None`, this indicates whether a Z2 symmetry reduction should be applied to resulting qubit operators that are computed based on mathematical symmetries that can be detected in the operator [de Keijzer].

- `name_solver`, the name of the solver, which defaults to `'NumPy exact solver'`.

- `solver`, either one of the solvers defined in Section *5.2.3, Setting up the VQE solver*, numpy_solver, vqe_ucc_solver, or vqe_tl_solver. It defaults to `NumPyMinimumEigensolver()`.

- `plot_bopes`, a Boolean, set to `True` to compute and plot the BOPES of the molecule.

- `perturbation_steps`, the points along the degrees of freedom to evaluate, in this case a distance in angstroms. It defaults to `np.linspace(-1, 1, 3)`.

- `pyqmc`, set to `True` by default to run the PyQMC Python module.

- n_ancillae, an integer that defaults to 3 that represents the number of ancillae qubits used by the run_qpe() function.

- num_time_slices, an integer that defaults to 1, which is number of **Trotterization** repetitions to make to improve the approximation accuracy. It's used by the Qiskit PauliTrotterEvolution class.

- loss, an optional array of floats that is generated by the callback function.

- label, a character string to be displayed by the plot_loss() function.

- target, a float to be displayed by the plot_loss() function.

- show, set to True by default to display intermediate results.

Here is the definition of the solve_ground_state() function:

```
def solve_ground_state(
    molecule,
    mapper ="Parity",
    num_electrons=None,
    num_molecular_orbitals=None,
    transformers=None,
    two_qubit_reduction=False,
    z2symmetry_reduction = "Auto",
    name_solver='NumPy exact solver',
    solver=NumPyMinimumEigensolver(),
    plot_bopes=False,
    perturbation_steps=np.linspace(-1, 1, 3),
    pyqmc=True,
    n_ancillae=3,
    num_time_slices=1,
    loss=[],
    label=None,
    target=None,
    show=True
):
```

We now present the code that is contained in the solve_ground_state() function.

We first define the electronic structure molecule driver by selecting the PySCF driver type and the basis set `sto3g` in which the molecular orbitals are to be expanded into. Here is the code:

```
# Defining the electronic structure molecule driver
driver = ElectronicStructureMoleculeDriver(molecule,
basis='sto3g', driver_type=ElectronicStructureDriverType.PYSCF)
```

Then, if both `num_electrons` and `num_molecular_orbitals` are specified, we call the `ActiveSpaceTransformer` function to split the computation into a classical and a quantum part:

```
# Splitting into classical and quantum
    if num_electrons != None and num_molecular_orbitals !=
None:
        split = ActiveSpaceTransformer(num_electrons=num_
electrons, num_molecular_orbitals=num_molecular_orbitals)
    else:
        split = None
```

Next, we create an `ElectronicStructureProblem` that produces the list of fermionic operators as follows:

```
# Defining a fermionic Hamiltonian operator
    if split != None:
        fermionic_hamiltonian =
ElectronicStructureProblem(driver, [split])
    elif transformers != None:
        fermionic_hamiltonian =
ElectronicStructureProblem(driver, transformers=transformers)
    else:
        fermionic_hamiltonian =
ElectronicStructureProblem(driver)
```

We then use the `second_q_ops()` method [Qiskit_Nat_3], which returns a list of second quantized operators: Hamiltonian operator, total particle number operator, total angular momentum operator, total magnetization operator, and if available, x, y, z dipole operators:

```
second_q_op = fermionic_hamiltonian.second_q_ops()
```

We get the particle number property of the molecule by calling the `particle_number()` function:

```
# Get particle number
particle_number = get_particle_number(fermionic_
hamiltonian, show=show)
```

If the input parameter `show` is set to `True`, we set truncation to `1000` with the `set_truncation(1000)` method and then we print the fermionic Hamiltonian operator of the molecule:

```
if show:
    # We set truncation to 1000 with the method set_
truncation(1000)
    second_q_op[0].set_truncation(1000)
    # then we print the first 20 terms of the fermionic
Hamiltonian operator of the molecule
    print("Fermionic Hamiltonian operator")
    print(second_q_op[0])
```

Next, we use the `fermion_to_qubit()` function defined in *Section 4.8, Constructing a qubit Hamiltonian with Qiskit Nature,* to convert a fermionic operator to a qubit operator:

```
# Use the function fermion_to_qubit() to convert a fermionic
operator to a qubit operator
    if show:
        print(" ")
    qubit_op, qubit_converter = fermion_to_qubit(fermionic_
hamiltonian, second_q_op, mapper=mapper, two_qubit_
reduction=two_qubit_reduction, z2symmetry_reduction=z2symmetry_
reduction, show=show)
```

Then we call the `run_PySCF()` function that we defined earlier to run the PySCF RHF method:

```
# Run the the PySCF RHF method
    if show:
        print(" ")
    conv, e = run_PySCF(molecule, pyqmc=pyqmc, show=show)
```

Then we call the `run_qpe()` function to perform a QPE and return the most likely eigenvalue of a Hamiltonian as an estimation of the electronic ground state energy:

```
# Run QPE
    eigv = run_qpe(particle_number, qubit_converter, qubit_
op, n_ancillae=n_ancillae, num_time_slices=num_time_slices,
show=True)
```

Next, we call the `run_vqe()` function defined earlier to solve for the ground state:

```
# Run VQE
if show:
  print(" ")
    ground_state = run_vqe(name_solver, fermionic_hamiltonian,
qubit_converter, solver, show=show)
```

If the `loss` parameter is not an empty array, we call the `plot_loss()` function to plot the evolution of the loss as a function of the number of iterations:

```
# Plot loss function
if loss != []:
    plot_loss(loss, label, target)
```

Next, if the `plot_bopes` parameter is set to `True`, we use the `BOPESSampler` Python class [Qiskit_Nat_6], which manages the process of varying the geometry and repeatedly calling the ground state solver, and then we get and plot the BOPES:

```
if plot_bopes:
    # Compute the potential energy surface as follows:
    energy_surface =
BOPESSampler(gss=GroundStateEigensolver(qubit_converter,
solver), bootstrap=False)
    # Set default to an empty dictionary instead of None:
    energy_surface._points_optparams = {}
    energy_surface_result = energy_surface.sample(fermionic_
hamiltonian, perturbation_steps)


    # Plot the energy as a function of atomic separation
    plot_energy_landscape(energy_surface_result)
```

Finally, we let the `solve_ground_state()` function return the following parameters to the caller:

- `fermionic hamiltonian`, the Fermionic Hamiltonian operator of the molecule
- `particle number`, the particle number property of the molecule
- `qubit_op`, the qubit Hamiltonian operator
- `qubit_converter`, either `JordanWignerMapper()`, `ParityMapper()`, or `BravyiKitaevMapper()`, which is the output of the `fermion_to_qubit()` function
- `ground_state`, the ground state of the molecule, if convergence has been achieved

with the following `return` statement:

```
return fermionic_hamiltonian, particle_number, qubit_op, qubit_
converter, ground_state
```

We now illustrate how to use the `solve_ground_state()` function with different molecules, different mappers, and different classical solvers.

5.2.1. Hydrogen molecule (H2)

We follow the process described in *Section 4.6.1, Constructing a fermionic Hamiltonian operator of the hydrogen molecule*. First, we define the geometry of the hydrogen molecule as follows:

```
hydrogen_molecule = Molecule(geometry=[['H', [0., 0., 0.]],
                                       ['H', [0., 0., 0.735]]],
                             charge=0, multiplicity=1)
```

We showed the particle number property of the hydrogen molecule in *Figure 4.9* in *Section 4.6.1, Constructing a fermionic Hamiltonian operator of the hydrogen molecule*, where we see four **spin orbitals (SOs)**, one α electron, and one β electron.

We showed the fermionic Hamiltonian operator of the hydrogen molecule in *Figure 4.13* in *Section 4.6.1, Constructing a fermionic Hamiltonian operator of the hydrogen molecule*.

Varying the hydrogen molecule

We specify the type of molecular variation, `Molecule.absolute_stretching`, as follows:

```
molecular_variation = Molecule.absolute_stretching
```

We specify that the first atom of the specified atom pair is moved closer to the second atom. The numbers refer to the index of the atom in the geometric definition list. Here is the code:

```
specific_molecular_variation = apply_variation(molecular_
variation, atom_pair=(1, 0))
```

We alter the original molecular definition as follows:

```
hydrogen_molecule_stretchable = Molecule(geometry=
                         [['H', [0., 0., 0.]],
                          ['H', [0., 0., 0.735]]],
                         charge=0, multiplicity=1,
                         degrees_of_freedom=[specific_
molecular_variation])
```

Now we proceed with solving for the ground state.

Solving for the ground state

We now run VQE using the NumPy exact minimum eigensolver:

```
H2_fermionic_hamiltonian, H2_particle_number, H2_qubit_op, H2_
qubit_converter, H2_ground_state = \
                solve_ground_state(hydrogen_molecule, mapper
="Parity",
                two_qubit_reduction=True, z2symmetry_
reduction=None,
                name_solver = 'NumPy exact solver', solver =
numpy_solver)
```

Figure 5.5 shows the results of the computation by the `run_PySCF()` and `run_QPE()` functions:

```
PySCF restricted HF (RHF) converged ground-state energy: -1.116998996754
Iteration, Energy, Error
0:          -1.1131 0.0097
1:          -1.1423 0.0083
2:          -1.1458 0.0073
3:          -1.1577 0.0087
PyQMC Monte Carlo converged ground-state energy: -1.157744954249, error: 0.0087

QPE initial Hartree Fock state
```

```
QPE computed electronic ground state energy (Hartree): -1.7934378610679784
```

Figure 5.5 – Ground-state of the H2 molecule with PySCF RHF, PyQMC Monte Carlo, and QPE

Figure 5.6 shows the result of the VQE computation:

```
Running the VQE using the NumPy exact solver
Elapsed time: 0:00:00.600964

=== GROUND STATE ENERGY ===

* Electronic ground state energy (Hartree): -1.857275030202
  - computed part:      -1.857275030202
~ Nuclear repulsion energy (Hartree): 0.719968994449
> Total ground state energy (Hartree): -1.137306035753

=== MEASURED OBSERVABLES ===

  0:  # Particles: 2.000 S: 0.000 S^2: 0.000 M: 0.000

=== DIPOLE MOMENTS ===

~ Nuclear dipole moment (a.u.): [0.0  0.0  1.3889487]
```

Figure 5.6 – Ground-state of the H2 molecule with VQE using the NumPy minimum eigensolver

Next, we run the VQE using the UCC factory ansatz [VQE_2] [VQE_3]:

```
H2_fermionic_hamiltonian, H2_particle_number, H2_qubit_op, H2_
qubit_converter, H2_ground_state = \
                solve_ground_state(hydrogen_molecule, mapper
="Parity",
                two_qubit_reduction=True, z2symmetry_
reduction=None,
                name_solver = 'Unitary Coupled Cluster (UCC)
factory ansatz', solver = vqe_ucc_solver)
```

Figure 5.7 shows the result:

```
Running the VQE using the Unitary Coupled Cluster (UCC) factory ansatz
Elapsed time: 0:00:01.581381

=== GROUND STATE ENERGY ===

* Electronic ground state energy (Hartree): -1.857275030145
  - computed part:      -1.857275030145
~ Nuclear repulsion energy (Hartree): 0.719968994449
> Total ground state energy (Hartree): -1.137306035696

=== MEASURED OBSERVABLES ===

  0:  # Particles: 2.000 S: 0.000 S^2: -0.000 M: 0.000

=== DIPOLE MOMENTS ===

~ Nuclear dipole moment (a.u.): [0.0  0.0  1.3889487]
```

Figure 5.7 – Ground-state of the H2 molecule with VQE using the UCC factory ansatz

Now we run the VQE using a heuristic ansatz, the Two-Local circuit with the default SLSQP optimizer [Panagiotis]:

```
H2_fermionic_hamiltonian, H2_particle_number, H2_qubit_op, H2_
qubit_converter, H2_ground_state = \
                solve_ground_state(hydrogen_molecule, mapper
="Parity",
                two_qubit_reduction=True, z2symmetry_
reduction=None,
                name_solver = 'Heuristic ansatz, the Two-
Local circuit with SLSQP',solver = vqe_tl_solver)
```

Figure 5.8 shows the result:

```
Running the VQE using the Heuristic ansatz, the Two-Local circuit with SLSQP
Elapsed time: 0:00:00.725543

=== GROUND STATE ENERGY ===

* Electronic ground state energy (Hartree): -1.857274878144
  - computed part:        -1.857274878144
~ Nuclear repulsion energy (Hartree): 0.719968994449
> Total ground state energy (Hartree): -1.137305883695

=== MEASURED OBSERVABLES ===

  0:  # Particles: 2.000 S: 0.000 S^2: -0.000 M: 0.000

=== DIPOLE MOMENTS ===

~ Nuclear dipole moment (a.u.): [0.0  0.0  1.3889487]
```

Figure 5.8 – Ground-state of the H2 molecule with VQE using the Two-Local circuit and SLSQP

We define the `qnspsa()` function as follows:

```
qnspsa_loss = []
ansatz = tl_circuit
fidelity = QNSPSA.get_fidelity(ansatz, quantum_instance,
expectation=PauliExpectation())
qnspsa = QNSPSA(fidelity, maxiter=200, learning_rate=0.01,
perturbation=0.7, callback=qnspsa_callback)
```

Here is the code that sets up the VQE using a heuristic ansatz and the QN-SPSA optimizer:

```
vqe_tl_QNSPSA_solver = VQE(ansatz=tl_circuit, optimizer=qnspsa,
                    quantum_instance=quantum_instance)
```

Now we call `solve_ground_state()` with the heuristic ansatz and the QN-SPSA optimizer:

```
H2_fermionic_hamiltonian, H2_particle_number, H2_qubit_op, H2_
qubit_converter, H2_ground_state = \
                solve_ground_state(hydrogen_molecule, mapper
="Parity",
```

```
                 two_qubit_reduction=True, z2symmetry_
reduction=None, loss=qnspsa_loss, label='QN-SPSA',
target=-1.857274810366,
                 name_solver='Two-Local circuit and the QN-
SPSA optimizer', solver=vqe_tl_QNSPSA_solver)
```

Figure 5.9 shows the result:

```
Running the VQE using the Two-Local circuit and the QN-SPSA optimizer
Elapsed time: 0:00:03.984191

=== GROUND STATE ENERGY ===

* Electronic ground state energy (Hartree): -1.857273494539
  - computed part:      -1.857273494539
~ Nuclear repulsion energy (Hartree): 0.719968994449
> Total ground state energy (Hartree): -1.13730450009

=== MEASURED OBSERVABLES ===

  0:  # Particles: 2.000 S: 0.000 S^2: 0.000 M: 0.000

=== DIPOLE MOMENTS ===

~ Nuclear dipole moment (a.u.): [0.0  0.0  1.3889487]

  0:
  * Electronic dipole moment (a.u.): [0.0  0.0  1.3889487]
    - computed part:      [0.0  0.0  1.3889487]
  > Dipole moment (a.u.): [0.0  0.0  0.0]  Total: 0.0
                 (debye): [0.0  0.0  0.0]  Total: 0.0
```

Figure 5.9 – Ground-state of the H2 molecule with VQE using the Two-Local circuit and QN-SPSA

Figure 5.10 shows the plot of the loss function of the QN-SPSA optimizer:

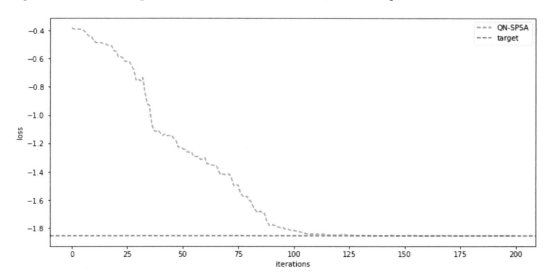

Figure 5.10 – Plot of the loss function of the VQE using the Two-Local circuit and QN-SPSA
for the H2 molecule

The table shown in *Figure 5.11* summarizes calculations obtained with the Python
packages PySCF RHF, PyQMC, and with the Qiskit Nature classes, VQE with NumPy
exact solver, SLSQP, QN-SPSA, and QPE.

Hydrogen Molecule	VQE					
Energy (Hartree)	NumPy	SLSQP	QN-SPSA	QPE	PySCF	PyQMC
Electronic ground state energy	-1.857	-1.857	-1.857	-1.793		
computed part	-1.857	-1.857	-1.857			
FreezeCoreTransformer extracted energy part						
Nuclear repulsion energy	0.720	0.720	0.720			
Total ground state energy	-1.137	-1.137	-1.137		-1.117	-1.162

Figure 5.11 – Table summarizing the calculations of the ground state energy obtained
with the H2 molecule

Figure 5.11 shows close agreement between the different calculations of the electronic ground state and the total ground state energies with the same qubit mapper called the `ParityMapper()` with `two_qubit_reduction=True`. The PyQMC method gives the lowest total energy -1.162 Ha and is the most accurate. It is consistent with the result -1.168 Ha shown in Ref. [Ebomwonyi].

Computing the BOPES

We now compute and plot the BOPES of the hydrogen molecule as follows:

```
perturbation_steps = np.linspace(-0.5, 2, 25) # 25 equally
spaced points from -0.5 to 2, inclusive.
H2_stretchable_fermionic_hamiltonian, H2_stretchable_particle_
number, H2_stretchable_qubit_op, H2_stretchable_qubit_
converter, H2_stretchable_ground_state = \
                solve_ground_state(hydrogen_molecule_
stretchable, mapper ="Parity",
                two_qubit_reduction=True, z2symmetry_
reduction=None,
                name_solver = 'NumPy exact solver', solver =
numpy_solver,
                plot_bopes = True, perturbation_
steps=perturbation_steps)
```

Figure 5.12 shows the plot of the BOPES of the hydrogen molecule:

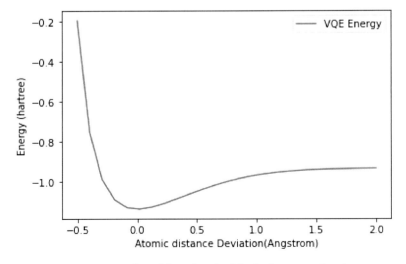

Figure 5.12 – Plot of the BOPES of the hydrogen molecule

5.2.2. Lithium hydride molecule

We follow the process described in *Section 4.6.2, Constructing a fermionic Hamiltonian operator of the lithium hydride molecule*. First, we define the geometry of the lithium hydride (LiH) molecule as follows:

```
LiH_molecule = Molecule(geometry=[['Li', [0., 0., 0.]],
                        ['H', [0., 0., 1.5474]]],
                 charge=0, multiplicity=1)
```

We showed the particle number property in *Figure 4.17* in *Section 4.6.2, Constructing a fermionic Hamiltonian operator of the lithium hydride molecule*, where we see six SOs, one α electron, and one β electron. We showed the fermionic Hamiltonian operator of the lithium hydride molecule in *Figure 4.20* in *Section 4.6.2, Constructing a fermionic Hamiltonian operator of the lithium hydride molecule*.

Varying the lithium hydride molecule

We alter the original molecular definition as follows:

```
LiH_molecule_stretchable = Molecule(geometry=[['Li', [0., 0.,
0.]],
                        ['H', [0., 0., 1.5474]]],
                 charge=0, multiplicity=1,
                 degrees_of_freedom=[specific_molecular_
variation])
reduction=True, z2symmetry_reduction="auto")
```

Solving for the ground state

We run VQE using the NumPy exact minimum eigensolver:

```
LiH_fermionic_hamiltonian, LiH_particle_number, LiH_qubit_op,
LiH_qubit_converter, LiH_ground_state = \
                solve_ground_state(LiH_molecule, mapper
="Parity",
                transformers=[FreezeCoreTransformer(freeze_
core=True, remove_orbitals=[4, 3])],
                two_qubit_reduction=True, z2symmetry_
reduction="auto",
                name_solver = 'NumPy exact solver', solver =
numpy_solver)
```

Figure 5.13 shows the result of the computation by the `run_PySCF()` and `run_QPE()` functions:

```
PySCF restricted HF (RHF) converged ground-state energy: -7.863113882796
Iteration, Energy, Error
0:          -7.7607 0.0897
1:          -8.2156 0.0594
2:          -8.0048 0.0450
3:          -8.1135 0.0355
PyQMC Monte Carlo converged ground-state energy: -8.113544597616, error: 0.0355

QPE initial Hartree Fock state
```

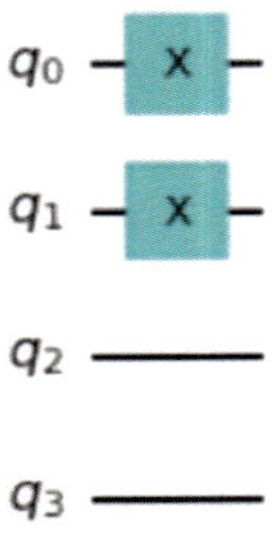

```
QPE computed electronic ground state energy (Hartree): -0.9611256062364334
```

Figure 5.13 – Ground state of the LiH molecule with PySCF RHF, PyQMC Monte Carlo, and QPE

Figure 5.14 shows the result of the VQE computation:

```
Running the VQE using the NumPy exact solver
Elapsed time: 0:00:01.413198

=== GROUND STATE ENERGY ===

* Electronic ground state energy (Hartree): -8.907396311316
  - computed part:       -1.088706015735
  - FreezeCoreTransformer extracted energy part: -7.818690295581
~ Nuclear repulsion energy (Hartree): 1.025934879643
> Total ground state energy (Hartree): -7.881461431673

=== MEASURED OBSERVABLES ===

  0:  # Particles: 2.000 S: 0.000 S^2: 0.000 M: 0.000

=== DIPOLE MOMENTS ===

~ Nuclear dipole moment (a.u.): [0.0  0.0  2.92416221]

  0:
  * Electronic dipole moment (a.u.): [0.0  0.0  4.76300889]
    - computed part:       [0.0  0.0  4.76695575]
    - FreezeCoreTransformer extracted energy part: [0.0  0.0  -0.00394686]
    > Dipole moment (a.u.): [0.0  0.0  -1.83884668]  Total: 1.83884668
```

Figure 5.14 – Ground state of the LiH molecule with VQE using the NumPy minimum eigensolver

We run the VQE using the Two-Local circuit and SLSQP:

```
LiH_fermionic_hamiltonian, LiH_particle_number, LiH_qubit_op,
LiH_qubit_converter, LiH_ground_state = \
                solve_ground_state(LiH_molecule, mapper
="Parity",
                transformers=[FreezeCoreTransformer(freeze_
core=True, remove_orbitals=[4, 3])],
                two_qubit_reduction=True, z2symmetry_
reduction="auto",
                name_solver = 'Heuristic ansatz, the Two-
Local circuit with SLSQP', solver = vqe_tl_solver)
```

Figure 5.15 shows the result:

```
Running the VQE using the Heuristic ansatz, the Two-Local circuit with SLSQP
Elapsed time: 0:00:02.406226

=== GROUND STATE ENERGY ===

* Electronic ground state energy (Hartree): -8.88904819163
  - computed part:      -1.070357896048
  - FreezeCoreTransformer extracted energy part: -7.818690295581
~ Nuclear repulsion energy (Hartree): 1.025934879643
> Total ground state energy (Hartree): -7.863113311986

=== MEASURED OBSERVABLES ===

  0:  # Particles: 2.000 S: 0.000 S^2: 0.000 M: -0.000

=== DIPOLE MOMENTS ===

~ Nuclear dipole moment (a.u.): [0.0  0.0  2.92416221]

  0:
  * Electronic dipole moment (a.u.): [0.0  0.0  4.83090809]
    - computed part:      [0.0  0.0  4.83485494]
    - FreezeCoreTransformer extracted energy part: [0.0  0.0  -0.00394686]
  > Dipole moment (a.u.): [0.0  0.0  -1.90674588]  Total: 1.90674588
              (debye): [0.0  0.0  -4.84646414]  Total: 4.84646414
```

Figure 5.15 – Ground state of the LiH molecule with VQE using the Two-Local circuit and SLSQP

We define the qnspsa() function as follows:

```
qnspsa_loss = []
ansatz = tl_circuit
fidelity = QNSPSA.get_fidelity(ansatz, quantum_instance,
expectation=PauliExpectation())
qnspsa = QNSPSA(fidelity, maxiter=500, learning_rate=0.01,
perturbation=0.7, callback=qnspsa_callback)
```

Here is the code that sets up the VQE using a heuristic ansatz and the QN-SPSA optimizer:

```
vqe_tl_QNSPSA_solver = VQE(ansatz=tl_circuit, optimizer=qnspsa,
                    quantum_instance=quantum_instance)
```

Now we call `solve_ground_state()` with the heuristic ansatz and the QN-SPSA optimizer:

```
LiH_fermionic_hamiltonian, LiH_particle_number, LiH_qubit_op,
LiH_qubit_converter, LiH_ground_state = \
                 solve_ground_state(LiH_molecule,
mapper="Parity",
                 transformers=[FreezeCoreTransformer(freeze_
core=True, remove_orbitals=[4, 3])],
                 two_qubit_reduction=True, z2symmetry_
reduction="auto", loss=qnspsa_loss, label='QN-SPSA',
target=-1.0703584,
                 name_solver='Two-Local circuit and the QN-
SPSA optimizer', solver=vqe_tl_QNSPSA_solver)
```

Figure 5.16 shows the result:

```
Running the VQE using the Two-Local circuit and the QN-SPSA optimizer
Elapsed time: 0:00:09.046346

=== GROUND STATE ENERGY ===

* Electronic ground state energy (Hartree): -8.889048751696
  - computed part:        -1.070358456115
  - FreezeCoreTransformer extracted energy part: -7.818690295581
~ Nuclear repulsion energy (Hartree): 1.025934879643
> Total ground state energy (Hartree): -7.863113872053

=== MEASURED OBSERVABLES ===

  0:  # Particles: 2.527 S: 0.724 S^2: 1.249 M: 0.264

=== DIPOLE MOMENTS ===

~ Nuclear dipole moment (a.u.): [0.0  0.0  2.92416221]

  0:
  * Electronic dipole moment (a.u.): [0.0  0.0  4.13661659]
    - computed part:        [0.0  0.0  4.14056345]
    - FreezeCoreTransformer extracted energy part: [0.0  0.0  -0.00394686]
  > Dipole moment (a.u.): [0.0  0.0  -1.21245438]  Total: 1.21245438
              (debye): [0.0  0.0  -3.08175134]  Total: 3.08175134
```

Figure 5.16 – Ground state of the LiH molecule with VQE using the Two-Local circuit and QN-SPSA

Figure 5.17 shows the plot of the loss function of the QN-SPSA optimizer:

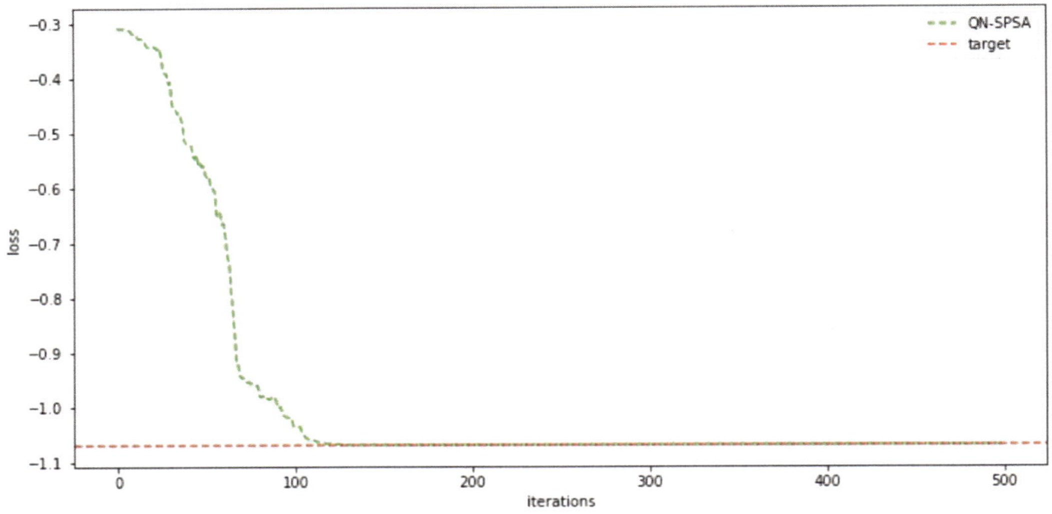

Figure 5.17 – Loss function of the VQE using the Two-Local circuit and QN-SPSA for the LiH molecule

The table shown in *Figure 5.18* summarizes calculations obtained with the Python packages PySCF RHF, PyQMC, and with the Qiskit Nature classes, VQE with the NumPy exact solver, SLSQP, QN-SPSA, and QPE:

Lithium hydride	VQE					
Energy (Hartree)	Numpy	SLSQP	QN-SPSA	QPE	PySCF	PyQMC
Electronic ground state energy	-8.907	-8.907	-8.889			
computed part	-1.089	-1.089	-1.070	-0.961		
FreezeCoreTransformer extracted energy part	-7.819	-7.819	-7.819			
Nuclear repulsion energy	1.026	1.026	1.026			
Total ground state energy	-7.881	-7.863	-7.863		-7.863	-8.102

Figure 5.18 – Table summarizing the calculations of the ground state energy obtained with the LiH molecule

Figure 5.18 shows close agreement between the different calculations of the electronic ground state and the total ground state energies. The PyQMC method gives the lowest total energy -8.102 Ha and is the most accurate. It is consistent with the result -8.07 Ha shown in Ref. [Adamowicz_3].

Computing the BOPES

We now compute and plot the BOPES of the lithium hydride molecule as follows:

```
perturbation_steps = np.linspace(-0.8, 0.8, 10) # 10 equally
spaced points from -0.8 to 0.8, inclusive.
LiH_stretchable_fermionic_hamiltonian, LiH_stretchable_
particle_number, LiH_stretchable_qubit_op, LiH_stretchable_
qubit_converter, LiH_stretchable_ground_state = \
                solve_ground_state(LiH_molecule_stretchable,
mapper ="Parity",
                transformers=[FreezeCoreTransformer(freeze_
core=True, remove_orbitals=[4, 3])],
                two_qubit_reduction=True, z2symmetry_
reduction="auto",
                name_solver='NumPy exact solver',
solver=numpy_solver,
                plot_bopes = True, perturbation_
steps=perturbation_steps)
```

Figure 5.19 shows the result:

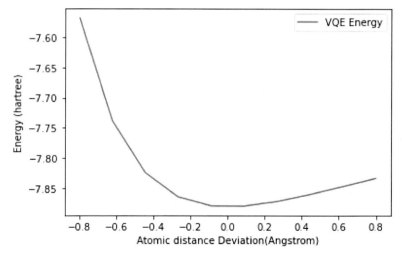

Figure 5.19 – Plot of the BOPES of the LiH molecule

5.2.3. Macro molecule

We now explore the HIV use case of the IBM Quantum Challenge Africa 2021, Quantum Chemistry for HIV [Africa21]. In their challenge they aimed to determine whether a toy model of an anti-retroviral molecule can bind with a toy model of a protease molecule. Since the anti-retroviral molecule has many atoms, it is approximated by using a single carbon atom. The toy model of the protease molecule is represented by a component of the formamide molecule (HCONH2); particularly it is the carbon-oxygen-nitrogen part of the formamide molecule. In short, the experiment is to determine whether a single carbon atom, can bind to the carbon-oxygen-nitrogen component of the formamide molecule. We will get the answer to the question posed by IBM by plotting the BOPES of a macro molecule, which is the formamide molecule plus the carbon atom.

First, we define the macro molecule with the ASE Atoms object [ASE_1]:

```
macro_ASE = Atoms('ONCHHHC', [(1.1280, 0.2091, 0.0000),
                              (-1.1878, 0.1791, 0.0000),
                              (0.0598, -0.3882, 0.0000),
                              (-1.3085, 1.1864, 0.0001),
                              (-2.0305, -0.3861, -0.0001),
                              (-0.0014, -1.4883, -0.0001),
                              (-0.1805, 1.3955, 0.0000)])
```

Then we display a 3D view of the molecule with the ASE viewer X3D for Jupyter notebooks [ASE2]:

```
view(macro_ASE, viewer='x3d')
```

Figure 5.20 shows the result. The nitrogen atom is depicted on the left side in blue, the oxygen atom on the right side in red, the carbon atoms in the middle in gray, and the three hydrogen atoms are the smallest ones in light gray. The carbon atom on the top is not bound to the other atoms.

Figure 5.20 – Macro molecule

We specify the type of molecular variation, `Molecule.absolute_stretching`, as follows:

```
molecular_variation = Molecule.absolute_stretching
```

We specify which atoms the variation applies to. The numbers refer to the index of the atom in the geometric definition list. The single carbon atom is moved closer to the nitrogen atom:

```
specific_molecular_variation = apply_variation(molecular_
variation, atom_pair=(6, 1))
```

We define the molecular geometry of the macro molecule with the Qiskit `Molecule` class as follows:

```
macromolecule = Molecule(geometry=
    [['O', [1.1280, 0.2091, 0.0000]],
    ['N', [-1.1878, 0.1791, 0.0000]],
    ['C', [0.0598, -0.3882, 0.0000]],
    ['H', [-1.3085, 1.1864, 0.0001]],
    ['H', [-2.0305, -0.3861, -0.0001]],
    ['H', [-0.0014, -1.4883, -0.0001]],
    ['C', [-0.1805, 1.3955, 0.0000]]],
    charge=0, multiplicity=1,
    degrees_of_freedom=[specific_molecular_variation])
```

Now we can solve for the ground state.

Solving for the ground state

We reduce the quantum workload by specifying that certain electrons should be treated with a quantum computing algorithm, while the remaining electrons should be classically approximated with the Qiskit `ActiveSpaceTransformer` class, which takes in two arguments:

- `num_electrons`, the number of electrons selected from the outermost electrons, counting inwards, to be treated with a quantum computing algorithm.

- `num_molecular_orbitals`, the number of orbitals to allow those electrons to roam over (around the so-called Fermi level). It determines how many qubits are needed.

We print the selection of parameters of the VQE run:

```
print("Macro molecule")
print("Using the ParityMapper with two_qubit_reduction=True to
eliminate two qubits")
print("Parameters ActiveSpaceTransformer(num_electrons=2, num_
molecular_orbitals=2)")
print("Setting z2symmetry_reduction=\"auto\"")
```

Here is the result:

```
Macro molecule
Using the ParityMapper with two_qubit_reduction=True to
eliminate two qubits
Parameters ActiveSpaceTransformer(num_electrons=2, num_
molecular_orbitals=2)
Setting z2symmetry_reduction="auto"
```

We then run the VQE using the NumPy exact minimum eigensolver:

```
macro_fermionic_hamiltonian, macro_particle_number, macro_
qubit_op, macro_qubit_converter, macro_ground_state = \
                solve_ground_state(macromolecule,
mapper="Parity",
                num_electrons=2, num_molecular_orbitals=2,
                two_qubit_reduction=True, z2symmetry_
reduction="auto",
                name_solver='NumPy exact solver',
solver=numpy_solver, pyqmc=False)
```

Figure 5.21 shows the first 20 terms of the fermionic operator of the macro molecule:

```
Number of alpha electrons: 1
Number of beta electrons: 1
Number of spin orbitals: 4
Fermionic Hamiltonian operator
Fermionic Operator
register length=4, number terms=26
  (0.026919512277770362+0j) * ( + 0 - 1 + 2 - 3 )
+ (-0.02691951227777035+0j) * ( +_0 -_1 -_2 +_3 )
+ (-0.05107681606888353+0j) * ( +_0 -_1 )
+ (-0.049268555669873024+0j) * ( +_0 -_1 +_3 -_3 )
+ (0.05107674503956438+0j) * ( +_0 -_1 +_2 -_2 )
+ (-0.02691951227777035+0j) * ( -_0 +_1 +_2 -_3 )
+ (0.02691951227777034+0j) * ( -_0 +_1 -_2 +_3 )
+ (0.05107681606888329+0j) * ( -_0 +_1 )
+ (0.04926855566987293+0j) * ( -_0 +_1 +_3 -_3 )
+ (-0.05107674503956443+0j) * ( -_0 +_1 +_2 -_2 )
+ (-0.05107681606888353+0j) * ( +_2 -_3 )
+ (0.05107681606888329+0j) * ( -_2 +_3 )
+ (-0.32719041369211194+0j) * ( +_3 -_3 )
+ (-0.7480534066333653+0j) * ( +_2 -_2 )
+ (0.27595173269703793+0j) * ( +_2 -_2 +_3 -_3 )
+ (-0.049268555669873024+0j) * ( +_1 -_1 +_2 -_3 )
+ (0.04926855566987293+0j) * ( +_1 -_1 -_2 +_3 )
+ (-0.32719041369211194+0j) * ( +_1 -_1 )
+ (0.519385326598636+0j) * ( +_1 -_1 +_3 -_3 )
+ (0.30287124497480833+0j) * ( +_1 -_1 +_2 -_2 )
+ (0.05107674503956438+0j) * ( +_0 -_0 +_2 -_3 )
+ (-0.0510767 ...
```

Figure 5.21 – First 20 terms of the fermionic Hamiltonian operator of the macro molecule

Figure 5.22 shows the qubit Hamiltonian operator for the outermost two electrons of the macro molecule obtained with the parity transformation. Only two qubits are needed as expected for a parity mapping of the fermionic Hamiltonian operator to the qubit Hamiltonian operator:

```
Qubit Hamiltonian operator
Parity transformation
Number of items in the Pauli list: 9
-0.641002818115757 * II
- 0.18731878004762867 * ZI
+ (0.1873187800476287+6.938893903907228e-18j) * IZ
- 0.13136975723491173 * ZZ
- 0.05017272138403766 * XI
+ (-0.05017265035471868+1.734723475976807e-18j) * XZ
+ (-0.05017272138403767-5.204170427930421e-18j) * IX
+ (0.05017265035471868-2.6020852139652106e-18j) * ZX
+ 0.02691951227777034 * XX
```

Figure 5.22 – Qubit Hamiltonian operator of the outermost two electrons of the macro molecule

Figure 5.23 shows the total ground state energy of the molecule computed by the PySCF RHF Python package and an estimation of the electronic ground state energy of the outermost two electrons of the molecule computed by the Qiskit Nature QPE class:

```
PySCF restricted HF (RHF) converged ground-state energy: -203.543863584135

QPE initial Hartree Fock state
```

```
QPE computed electronic ground state energy (Hartree): -0.8244072113871544
```

Figure 5.23 – Total and electronic ground state energy of the macro molecule by PySCF and QPE respectively

Figure 5.24 shows the result of VQE computation:

```
Running the VQE using the NumPy exact solver
Elapsed time: 0:39:53.346774

=== GROUND STATE ENERGY ===

* Electronic ground state energy (Hartree): -318.206223981746
  - computed part:        -0.885465166125
  - ActiveSpaceTransformer extracted energy part: -317.320758815622
~ Nuclear repulsion energy (Hartree): 114.661165852466
> Total ground state energy (Hartree): -203.545058129281

=== MEASURED OBSERVABLES ===

  0:  # Particles: 2.000 S: 0.000 S^2: 0.000 M: 0.000

=== DIPOLE MOMENTS ===

~ Nuclear dipole moment (a.u.): [-6.34040909  15.65127868  -0.00018897]

  0:
  * Electronic dipole moment (a.u.): [-5.12026581  16.91295797  -0.00024077]
    - computed part:        [-0.70824787  4.63305797  -0.00017904]
    - ActiveSpaceTransformer extracted energy part: [-4.41201795  12.2799  -0.00006172]
  > Dipole moment (a.u.): [-1.22014328  -1.26167929  0.0000518]  Total: 1.75515933
              (debye): [-3.10129458  -3.20686858  0.00013165]  Total: 4.4611696
```

Figure 5.24 – Ground state of macro molecule using the NumPy exact minimum eigensolver

The electronic ground state energy of the outermost two electrons of the macromolecule computed by the QPE, -0.824 (Hartree), and by the VQE, -0.885, Qiskit Nature classes are in good agreement.

The total ground state energy of the macro molecule computed by the PySCF RHF Python package, -203.54386 (Hartree), and by the Qiskit Nature VQE class, -203.54505, are in good agreement.

Computing the BOPES

We now compute and plot the BOPES of the macro molecule as follows:

```python
perturbation_steps = np.linspace(-0.5, 3, 10) # 10 equally
spaced points from -0.5 to 3, inclusive.
macro_fermionic_hamiltonian, macro_particle_number, macro_
qubit_op, macro_qubit_converter, macro_ground_state = \
            solve_ground_state(macromolecule, mapper
="Parity",
            num_electrons=2, num_molecular_orbitals=2,
```

```
            two_qubit_reduction=True, z2symmetry_
reduction="auto",
            name_solver='NumPy exact solver',
solver=numpy_solver, pyqmc=False,
            plot_bopes=True, perturbation_
steps=perturbation_steps)
```

Figure 5.25 shows the result:

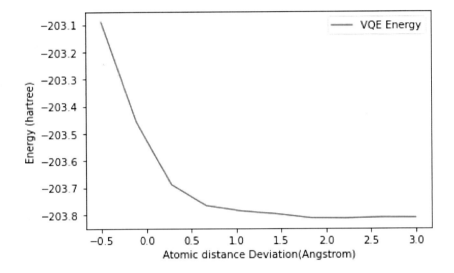

Figure 5.25 – Plot of the BOPES of the macro molecule

The plot of the BOPES of the macro molecule shows no clear minimum for any separation. We conclude that there is no binding of the single carbon atom to the toy protease molecule of formamide.

Summary

In this chapter, we have introduced classical and hybrid classical-quantum variational methods to find the lowest energy eigenvalue for a quantum system and their implementation with a classical PyQMC variational Monte Carlo Python package, which interoperates with the PySCF, and Qiskit Nature using the STO-3G basis with the Python-based PySCF driver.

We have illustrated these methods, solving for the ground state and plotting the BOPES of the hydrogen molecule, the lithium hydride molecule, and the macro molecule.

The results we obtained with Qiskit Nature VQE and QPE are in good agreement with those obtained with the PyQMC and PySCF RHF packages for several combinations of fermionic-to-qubit Hamiltonian mappers and classical gradient descent solvers and by reducing the quantum workload to the outermost two electrons of the formamide molecule. We hope these results will encourage the reader to replay these experiments with different choices of solvers and with other molecules.

Questions

Please test your understanding of the concepts presented in this chapter with the corresponding Google Colab notebook:

1.   Does the variational theorem apply to excited states?

2.   True or False: The Metropolis-Hastings method is a way to approximate integration over spatial coordinates.

3.   True or False: VQE is only a quantum computing algorithm and does not require the use of classical computing.

Answers

1.   Yes

2.   True

3.   False

References

[ASE_0] Atomic Simulation Environment (ASE), https://wiki.fysik.dtu.dk/ase/index.html

[ASE_1] ASE, The Atoms object, https://wiki.fysik.dtu.dk/ase/ase/atoms.html

[ASE_2] ASE Visualization, https://wiki.fysik.dtu.dk/ase/ase/visualize/visualize.html#module-ase.visualize

[Adamowicz_3] Tung WC, Pavanello M, Adamowicz L., Very accurate potential energy curve of the LiH molecule. TABLE I. Comparison of the convergence of the BO energy, in Eh, for the ground state of LiH molecule at R = 3.015 bohrs, J Chem Phys. 2011 Feb 14;134(6):064117. doi: 10.1063/1.3554211, https://doi.org/10.1063/1.3554211

[Africa21] IBM Quantum Challenge Africa 2021, https://github.com/qiskit-community/ibm-quantum-challenge-africa-2021

[Burg] Vera von Burg, Guang Hao Low, Thomas Häner, Damian S. Steiger, Markus Reiher, Martin Roetteler, Matthias Troyer, Quantum computing enhanced computational catalysis, 3 Mar 2021, 10.1103/PhysRevResearch.3.033055, https://arxiv.org/abs/2007.14460

[Bushmanov] Sergey Bushmanov, How to plot a histogram using Matplotlib in Python with a list of data?, Stack Overflow, https://stackoverflow.com/questions/33203645/how-to-plot-a-histogram-using-matplotlib-in-python-with-a-list-of-data

[Cao] Yudong Cao, Jonathan Romero, Jonathan P. Olson, Matthias Degroote, Peter D. Johnson, Mária Kieferová, Ian D. Kivlichan, Tim Menke, Borja Peropadre, Nicolas P. D. Sawaya, Sukin Sim, Libor Veis, Alán Aspuru-Guzik, Quantum Chemistry in the Age of Quantum Computing, Chem. Rev. 2019, 119, 19, 10856–10915, Aug 30, 2019, https://doi.org/10.1021/acs.chemrev.8b00803

[Chen] Sija Chen, Quantum Monte Carlo Methods, Maplesoft, https://fr.maplesoft.com/Applications/Detail.aspx?id=154748

[Dagrada] Mario Dagrada, Improved quantum Monte Carlo simulations: from open to extended systems, Materials Science [cond-mat.mtrl-sci]. Université Pierre et Marie Curie - Paris VI; Universidad Nacional de San Martín, 2016. English. ⟨NNT: 2016PA066349⟩. ⟨tel-01478313⟩, https://tel.archives-ouvertes.fr/tel-01478313/document

[Ebomwonyi] Ebomwonyi, Osarodion, A Quantum Monte Carlo Calculation of the Ground State Energy for the Hydrogen Molecule Using the CASINO Code, 2013, Table 3.1: Comparative analysis of the ground state energies for the hydrogen molecule by different researchers, https://www.semanticscholar.org/paper/A-Quantum-Monte-Carlo-Calculation-of-the-Ground-for-Ebomwonyi/5316eb86f39cf4fa0a8fd06d136aac4db1105ad4

[Freeman] Freedman–Diaconis rule, Wikipedia, https://en.wikipedia.org/wiki/Freedman%E2%80%93Diaconis_rule

[Gorelov] Vitaly Gorelov, Quantum Monte Carlo methods for electronic structure calculations: application to hydrogen at extreme conditions, 1.4.1 Variational Monte Carlo (VMC), https://tel.archives-ouvertes.fr/tel-03045954/document

[Grok] Grok the Bloch Sphere, https://javafxpert.github.io/grok-bloch/

[H5py] Quick Start Guide, https://docs.h5py.org/en/stable/quick.html

[IBM_CEO] IBM CEO: Quantum computing will take off 'like a rocket ship' this decade, Fast Company, Sept 28, 2021., https://www.fastcompany.com/90680174/ibm-ceo-quantum-computing-will-take-off-like-a-rocket-ship-this-decade

[IBM_comp1] Welcome to IBM Quantum Composer, https://quantum-computing.ibm.com/composer/docs/iqx/

[IBM_comp2] IBM Quantum Composer, https://quantum-computing.ibm.com/composer/files/new

[Lolur] Lolur, Phalgun, Magnus Rahm, Marcus Skogh, Laura García-Álvarez and Göran Wendin, Benchmarking the Variational Quantum Eigensolver through Simulation of the Ground State Energy of Prebiotic Molecules on High-Performance Computers, arXiv:2010.13578v2 [quant-ph], 5 Jan 2021, https://arxiv.org/pdf/2010.13578.pdf

[NumPy] NumPy: the absolute basics for beginners, https://numpy.org/doc/stable/user/absolute_beginners.html

[Panagiotis] Panagiotis Kl. Barkoutsos, Jerome F. Gonthier, Igor Sokolov, Nikolaj Moll, Gian Salis, Andreas Fuhrer, Marc Ganzhorn, Daniel J. Egger, Matthias Troyer, Antonio Mezzacapo, Stefan Filipp, Ivano Tavernelli, Quantum algorithms for electronic structure calculations: Particle-hole Hamiltonian and optimized wave-function expansions, Phys. Rev. A 98, 022322 – Published 20 August 2018, DOI: 10.1103/PhysRevA.98.022322, https://link.aps.org/doi/10.1103/PhysRevA.98.022322, https://arxiv.org/abs/1805.04340

[Pease] Christopher Pease, An Overview of Monte Carlo Methods, Towards Data Science, https://towardsdatascience.com/an-overview-of-monte-carlo-methods-675384eb1694

[PyQMC] PyQMC, a python module that implements real-space quantum Monte Carlo techniques, https://github.com/WagnerGroup/pyqmc

[PySCF] The Python-based Simulations of Chemistry Framework (PySCF), https://pyscf.org/

[Qiskit] Qiskit, `https://qiskit.org/`

[Qiskit_2021_Lab4] Julien Gacon, Lab 4: Introduction to Training Quantum Circuits, Qiskit Summer School 2021, `https://learn.qiskit.org/summer-school/2021/lab4-introduction-training-quantum-circuits`

[Qiskit_Nat_0] Qiskit_Nature, `https://github.com/Qiskit/qiskit-nature/blob/main/README.md`

[Qiskit_Nat_3] ElectronicStructureProblem.second_q_ops, `https://qiskit.org/documentation/nature/stubs/qiskit_nature.problems.second_quantization.ElectronicStructureProblem.second_q_ops.html`

[Qiskit_Nat_4] QubitConverter, `https://qiskit.org/documentation/nature/stubs/qiskit_nature.converters.second_quantization.QubitConverter.html`

[Qiskit_Nat_5] Qiskit Nature Tutorials, Electronic structure, `https://qiskit.org/documentation/nature/tutorials/01_electronic_structure.html`

[Qiskit_Nat_6] Qiskit Nature Tutorials, Sampling the potential energy surface, `https://qiskit.org/documentation/nature/_modules/qiskit_nature/algorithms/pes_samplers/bopes_sampler.html`

[Qiskit_Nature] Introducing Qiskit Nature, Qiskit, Medium, April 6, 2021, `https://medium.com/qiskit/introducing-qiskit-nature-cb9e588bb004`

[QuTiP] QuTiP, Plotting on the Bloch Sphere, `https://qutip.org/docs/latest/guide/guide-bloch.html`

[SciPy] Statistical functions (scipy.stats), `https://docs.scipy.org/doc/scipy/getting_started.html`

[Stephens] Matthew Stephens, The Metropolis Hastings Algorithm, `https://stephens999.github.io/fiveMinuteStats/MH_intro.html`

[Toulouse] Julien Toulouse, Introduction to quantum chemistry, Jan 20, 2021, `https://www.lct.jussieu.fr/pagesperso/toulouse/enseignement/introduction_qc.pdf`

[Tamir] Abraham Tamir, Applications of Markov Chains in Chemical Engineering, Elsevier, 1998, 9780080527390, 0080527396, `https://www.google.fr/books/edition/Applications_of_Markov_Chains_in_Chemica/X0ivOmHYPoYC`

[Toulouse_1] Julien Toulouse, Quantum Monte Carlo wave functions and their optimization for quantum chemistry, CEA Saclay, SPhN Orme des Merisiers, April 2015, `https://www.lct.jussieu.fr/pagesperso/toulouse/presentations/presentation_saclay_15.pdf`

[Troyer] Matthias Troyer, Matthias Troyer: Achieving Practical Quantum Advantage in Chemistry Simulations, QuCQC 2021, `https://www.youtube.com/watch?v=2MsfbPlKgyI`

[VQE_1] Peruzzo, A., McClean, J., Shadbolt, P. et al., A variational eigenvalue solver on a photonic quantum processor, Nat Commun 5, 4213 (2014), `https://doi.org/10.1038/ncomms5213`

[VQE_2] Qiskit Nature, Ground state solvers, `https://qiskit.org/documentation/nature/tutorials/03_ground_state_solvers.html`

[VQE_3] Hardware-efficient variational quantum eigensolver for small molecules and quantum magnets, Nature 549, 242–246 (2017), `https://doi.org/10.1038/nature23879`

[VQE_4] Running VQE on a Statevector Simulator, `https://qiskit.org/textbook/ch-applications/vqe-molecules.html#Running-VQE-on-a-Statevector-Simulator`

6
Beyond Born-Oppenheimer

"The first principle is that you must not fool yourself – and you are the easiest person to fool."

– Richard Feynman

"Scientific progress is measured in units of courage, not intelligence."

– Paul Dirac

Figure 6.1 – Dr. Keeper Sharkey imagining molecular vibrations of a diatomic molecule [authors]

Determining molecular structure and vibrational spectra computationally are two essential goals of modern computational chemistry that have applications in many areas, from astrochemistry to biochemistry and climate change mitigation. The computational complexity grows exponentially when the number of atoms and/or identical particles increases linearly. There is additional complexity associated when there are significant couplings between rotational and vibrational degrees of freedom, and at high energy states, including near the dissociation and ionization limit. Innovative computational methods and new quantum computing technology are actively being developed to address these hurdles [Sawaya].

Computational methods that can achieve next-generation accuracy will go beyond the standard approximations present in this book. This chapter focuses on introducing how non-Born-Oppenheimer (non-BO) calculations include the effects that are needed to make better predictions of chemical states at above ground state vibrations [Adamowicz_1][Adamowicz_2]. Other beyond BO-type approaches are being pursued to overcome the limitations of the BO approximation, however are not fully non-BO, such as pre-BO [Schiffer_1] [Schiffer_2][Mátyus][D4.1 VA Beta].

As an example of the benefits of utilizing vibrational states, we point to a team of physicists at the Massachussets Institute of Technology's Research Laboratory of Electronics that has demonstrated a new quantum register of fermion pairs where information is stored in the vibrational motion of atom pairs held in a superposition of two vibrational states [Hartke]. The common and relative motion of each atom pair is protected by exchange symmetry, enabling long-lived and robust motional coherence. They say *"Thus fermion anti-symmetry and strong interactions, the core challenges for classical computations of many-fermion behavior, may offer decisive solutions for protecting and processing quantum information."* Their achievement paves the way to building programmable quantum simulators of many-fermion behavior and digital computing using fermion pairs.

In this chapter, we will cover the following topics:

- *Section 6.1, Non-Born-Oppenheimer molecular Hamiltonian*

- *Section 6.2, Vibrational frequency analysis calculations*

- *Section 6.3, Vibrational spectra for ortho-para isomerization of hydrogen molecule*

Technical requirements

A companion Jupyter notebook for this chapter can be downloaded from GitHub at `https://github.com/PacktPublishing/Quantum-Chemistry-and-Computing-for-the-Curious`, which has been tested in the Google Colab environment, which is free and runs entirely in the cloud, and in the IBM Quantum Lab environment. Please refer to *Appendix B – Leveraging Jupyter Notebooks in the Cloud*, for more information. The companion Jupyter notebook automatically installs the following list of libraries:

- NumPy [NumPy], an open-source Python library that is used in almost every field of science and engineering
- SciPy [SciPy], a free and open-source Python library used for scientific computing and technical computing

The companion Jupyter notebook does not include the installation of the Psi4 free and open source software for high-throughput quantum chemistry [Psi4_0], which we used to perform a simple calculation of the vibrational frequency analysis of the carbon dioxide (CO_2) molecule. We refer the reader interested in installing this package to the "Get Started with Psi4" [Psi4_1] documentation and to the article Ref. [Psi4_3].

Installing NumPy, SimPy, and math modules

Install NumPy with the following command:

```
pip install numpy
```

Install SciPy with the following command:

```
pip install scipy
```

Import NumPy with the following command:

```
import numpy as np
```

Import Matplotlib, a comprehensive library for creating static, animated, and interactive visualizations in Python with the following command:

```
import matplotlib.pyplot as plt
```

Import the SciPy special Hermite polynomials with the following command:

```
from scipy.special import hermite
```

Import the math `factorial` function with the following command:

```
from math import factorial
```

6.1. Non-Born-Oppenheimer molecular Hamiltonian

Recall from *Section 4.1, Born-Oppenheimer approximation*, the expression of the Hamiltonian in the laboratory frame coordinates $\hat{H}_{LAB}$:

$$\hat{H}_{LAB} = -\frac{1}{2}\sum_{p=1}^{N} \nabla^2_{r_p} - \sum_{A=1}^{M} \frac{1}{2M_A} \nabla^2_{R_A} - \sum_{p=1}^{N}\sum_{A=1}^{M} \frac{Z_A}{r_{pA}} + \sum_{q>p=1}^{N} \frac{1}{r_{pq}} + \sum_{B>A=1}^{M} \frac{Z_A Z_B}{R_{AB}}$$

where in atomic units, the mass of the electron and the reduced Planck constant ($\hbar$) are set to the value 1. The LAB Hamiltonian comprises the sum of the kinetic energy of all particles and the potential energy between all particles with the following definitions:

- $\nabla^2_{r_p}$ and $\nabla^2_{R_A}$ are the second derivative operator with respect to the position coordinates for electrons and nuclei, that is, $\nabla^2_{R_A} = \frac{\partial^2}{\partial^2_{x_A}} + \frac{\partial^2}{\partial^2_{y_A}} + \frac{\partial^2}{\partial^2_{z_A}}$ and likewise for the pth electron.

- $r_{pq} = |r_p - r_q|$, $r_{pA} = |r_p - R_A|$, and $R_{AB} = |R_A - R_B|$ are the distances between electrons p and q, electron p and nucleus A, and nuclei A and B determined by the Euclidean norm.

The list of the operators of the LAB Hamiltonian has been presented in *Figure 4.3*.

In the LAB Hamiltonian, the energy of the molecular system is continuous, not discrete. The **center-of-mass (COM)** motion does not yield any change to the energy of the internal states of the system and can be factored out. The internal states are quantized and invariant to translations. These states are not affected by translational and rotational motions in free space. The nuclei can still move around the COM through vibrations and internal rotations.

In the BO approximation, we assume that the motions of the nuclei are uncoupled from the motions of the electrons, that is, a product of nuclear equations (rotational and vibrational) and electronic equations:

$$|\Psi_{total}(\boldsymbol{r}, \boldsymbol{s}, \boldsymbol{R})\rangle = \Psi_{rotational}(\boldsymbol{R})\Psi_{vibrational}(\boldsymbol{R})\Psi_{elec}(\boldsymbol{r}, \boldsymbol{s}; \boldsymbol{R})$$

where $\boldsymbol{R} = \{\boldsymbol{R}_A, \boldsymbol{R}_B, ..., \boldsymbol{R}_M\}$ are the nuclear coordinates, $\boldsymbol{r} = \{\boldsymbol{r}_p, \boldsymbol{r}_i, \boldsymbol{r}_j, ..., \boldsymbol{r}_N\}$ are the electron coordinates, $\boldsymbol{s} = \{\boldsymbol{s}_p, \boldsymbol{s}_i, \boldsymbol{s}_j, ..., \boldsymbol{s}_N\}$ are the spin coordinates, and the electronic wave function ($\Psi_{elec}(\boldsymbol{r}, \boldsymbol{s}; \boldsymbol{R})$) is conditioned on the nuclear coordinates ($\boldsymbol{R}$).

In the non-BO method, the total wave function is still separable in terms of rotations and vibrations and electronic energy levels; however, all energy levels are dependent on the variables for electrons and nuclei, spatial and spin.

$$|\Psi_{total}(\boldsymbol{r}, \boldsymbol{s}, \boldsymbol{R}, \boldsymbol{S})\rangle = \Psi_{rotational}(\boldsymbol{r}, \boldsymbol{s}, \boldsymbol{R}, \boldsymbol{S})\Psi_{vibrational}(\boldsymbol{r}, \boldsymbol{s}, \boldsymbol{R}, \boldsymbol{S})\Psi_{elec}(\boldsymbol{r}, \boldsymbol{s}, \boldsymbol{R}, \boldsymbol{S})$$

In the BO approximation, solving for only the electronic equation with fixed position of the nuclei can be iterated to account for the vibrations and internal rotations of the nuclei. For each iteration, the nuclei of the atoms are fixed in space and not moving and can be thought of as a violation of the Heisenberg uncertainty principle, introduced in *Section 1.4, Light and energy*. The more you know exactly where a particle is, the less you know about its momentum. In general, the internal coordinate system can be placed at the heaviest atom in the molecule or at the COM. This approximation has limitations that we will cover, specifically through vibrational spectra energy.

The non-BO method presented in this chapter exploits the effectiveness and predictive power of all-particle explicit correlation utilizing **explicitly correlated Gaussian basis functions (ECGs)** [Adamowicz_4] to describe atomic and molecular phenomena [Sharkey]. This method can be used to model a small number of particles, that is, three nuclei or no more than seven electrons, as well as determining highly accurate ionization energies of atoms, Rydberg states, and rotational and vibrationally excited states to an arbitrary level [Sharkey]. Furthermore, this method is very amenable to an effective parallelization.

Internal Hamiltonian operator

We consider a general non-relativistic atomic system consisting of N particles, that is, $n = N-1$ electrons and a nucleus. Converting from the laboratory frame Hamiltonian to the internal Hamiltonian involves rigorously separating the COM motion from the laboratory frame. We define an internal Cartesian coordinate frame (CCF) where all particles are treated on equal footing and all particles are allowed to roam space freely without any constraints. The change of coordinates from laboratory CCF to internal CCF is illustrated in *Figure 6.2*.

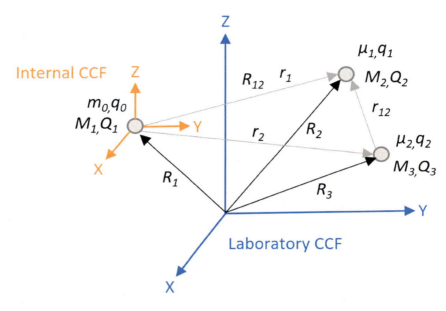

Figure 6.2 – Internal CCF and laboratory CCF

The resulting internal Hamiltonian is used to calculate bound states of the system:

$$\widehat{H}_{Int} = -\frac{1}{2}\left(\sum_{i=1}^{n}\frac{1}{\mu_i}\nabla_{r_i}^2 + \sum_{i\neq j}^{n}\frac{1}{m_0}\nabla'_{r_i}\nabla_{r_j}\right) + \sum_{i=1}^{n}\frac{q_0 q_i}{r_i} + \sum_{i>j=1}^{n}\frac{q_i q_j}{r_{ij}}$$

For clarity, we list the terms of the internal Hamiltonian in *Figure 6.3*.

Term	Description
m_0	Mass of the central particle.
$\mu_i = m_0 m_i / (m_0 + m_i)$	Reduced mass of the pseudo particle i.
q_i	Charge of the pseudo particle i.
$r_{ij} = \lvert \mathbf{r}_j - \mathbf{r}_i \rvert$	Relative distance between the pseudo particles i and j.
$\displaystyle\sum_{i=1}^{n} \frac{1}{\mu_i} \nabla_{\mathbf{r}_i}^2$	Electronic kinetic energy operator of each pseudo particle.
$\displaystyle\sum_{i \neq j}^{n} \frac{1}{m_0} \nabla'_{\mathbf{r}_i} \nabla_{\mathbf{r}_j}$	Kinetic energy coupling terms, also known as the mass polarization terms (or BO terms).
$\displaystyle\sum_{i=1}^{n} \frac{q_0 q_i}{r_i}$	Coulomb interactions between pseudo particles and the central particle.
$\displaystyle\sum_{i>j=1}^{n} \frac{q_i q_j}{r_{ij}}$	Coulomb interactions between pair pseudo particles.

Figure 6.3 – Terms of the internal Hamiltonian operator for a molecule

We now present the all-particle explicit correlation approach.

Explicitly correlated all-particle Gaussian functions

The ECGs that are used in the non-BO method are shown in *Figure 6.4*.

Term	Description
$\Psi_{total}(\mathbf{r}, \mathbf{s}, \mathbf{R}) = \displaystyle\sum_{k}^{K} c_k \widehat{Y}_k \, \phi_k(\mathbf{r}) \chi_k(\mathbf{s})$	The wave function of the system is expanded with explicitly correlated all-particle Gaussian functions.
c_k	Linear expansion coefficients.
$\widehat{Y}_k$	Symmetry projection constructed from the Young tableaux introduced in *Section 2.1.3, General formulation of the Pauli exclusion principle*.
$\phi_k(\mathbf{r}) \equiv \exp[-\mathbf{r}'\mathbf{A}_k\mathbf{r}]$	ECG basis functions. They are rotationally invariant.
$\mathbf{A}_k = A_k \otimes I_3$	$A_k = L_k L_k'$ is the Cholesky factorization to ensure positive definiteness, and I_3 is the 3×3 identity matrix.
$\chi_k(\mathbf{s})$	Spin function.

Figure 6.4 – ECGs

The non-BO method uses the variational principle that was introduced in *Chapter 5, Variational Quantum Eigensolver (VQE) Algorithm,* and specifically includes an energy minimization procedure.

Energy minimization

To obtain the eigenvalues of the LAB Hamiltonian $\widehat{H}_{LAB}$, we use the Rayleigh-Ritz variational scheme based on the minimization of the Rayleigh quotient:

$$\varepsilon(a, c) = \min_{(a,c)} \frac{c'H(a)c}{c'S(a)c}$$

where $H(a)$ and $S(a)$ are the Hamiltonian and overlap $K \times K$ matrices, respectively. $H(a)$ and $S(a)$ are functions of the nonlinear parameters contained in the basis-set of ECGs. We write a for the set of these parameters and c for the vector of the linear expansion coefficients of the wave function in terms of basis functions.

We derive and implement the analytic gradient of the energy with respect to the nonlinear parameters of the Gaussians, starting with the secular equation $(H - \varepsilon S)C = 0$:

$$d(H - \varepsilon S)C = (dH)c - (d\varepsilon)Sc - \varepsilon(dS)c + (H - \varepsilon S)dc$$

Multiplying this equation by c' from the left, we obtain the well-known Hellmann-Feynman theorem:

$$d\varepsilon = c'(dH - \varepsilon dS)c$$

To obtain this expression, we utilize the secular equation, and we assume that the wave function is normalized, that is, $c'Sc = 1$. The expression for $d\varepsilon$ involves dH and dS, which depend on the first derivatives of the Hamiltonian and overlap integrals with respect to the Gaussian nonlinear parameters.

The method employs explicitly correlated all-particle Gaussian functions for expanding the wave function of the system. The nonlinear parameters of the Gaussians are variationally optimized with an approach employing an analytical energy gradient determined with respect to these parameters.

6.2. Vibrational frequency analysis calculations

Within the BO approximation, the total energy of a molecule is the sum of the electronic, vibrational, and rotational energy:

$$E_{total} = E_{rotational} + E_{vibrational} + E_{elec}$$

Molecular vibrations can be modeled like the motion of particles connected by springs, representing atoms connected by chemical bonds of variable lengths. In the harmonic oscillator approximation, the force required to extend the spring is proportional to the extension (Hooke's law). When the vibrational energy is high, the harmonic oscillator approximation is no longer valid, and neither is the concept of normal mode. We now consider the simple case of a diatomic molecule.

Modeling the vibrational-rotational levels of a diatomic molecule

The rotational energies of a diatomic molecule are represented by a series of discrete values:

$$E_{rotational} = \frac{J(J+1)\hbar^2}{2MR_e^2}$$

where:

- J is the angular momentum quantum number.
- R_e is the equilibrium distance.
- $\hbar$ is the reduced Planck constant.
- M is the reduced mass of the two atoms, $M = \frac{M_A M_B}{M_A + M_B}$.

For a non-rotating diatomic molecule, the rotational quantum number J is zero. The potential energy $E_{pot}(x)$ can be approximated by a quadratic function of the displacement x around the equilibrium position $x = 0$, which corresponds to the equilibrium distance R_e between the two nuclei:

$$E_{pot}(x) = \frac{1}{2}kx^2$$

where k is the restoring force constant (Hooke's law). For such a parabolic potential, the vibrating molecule is a quantum harmonic oscillator. The energy levels are a simple function of the integer vibrational quantum number v:

$$E_v = \hbar\omega\left(v + \frac{1}{2}\right)$$

where $\omega = \sqrt{k/M}$ depends on the constant k and the reduced mass M of the two atoms.

The normalized wave functions of the quantum harmonic oscillator are:

$$\psi_v(x) = \frac{1}{\sqrt{2^v\, v!}} \cdot \left(\frac{M\omega}{\pi\hbar}\right)^{1/4} \cdot H_v\left(\sqrt{\frac{M\omega}{\hbar}}\, x\right) \cdot e^{-M\omega x^2/2\hbar}$$

The Hermite polynomial of order v is defined by the generation equation:

$$H_v(z) = (-1)^v\left(e^{z^2}\right)\frac{d^v}{dz^v}\left(e^{-z^2}\right)$$

Hermite polynomials are computed using the following recursion relation:

$$z\,H_v(z) = z\,H_{v-1}(z) + \frac{1}{2}\,H_{v+1}(z)$$

The first three Hermite polynomials are:

$$H_0(z) = 1$$
$$H_1(z) = 2z$$
$$H_2(z) = 4z^2 - 2$$

We define the N(v) function, which computes the normalization factor:

$$N(v) = \left(\sqrt{\pi}\,2^v\,v!\right)^{-1/2}$$

```
def N(v):
    return 1./np.sqrt(np.sqrt(np.pi)*2**v*factorial(v))
```

We define the Psi(v, x) function, which uses the special hermite() SciPy function and computes a function with the same form as the normalized wave functions of the quantum harmonic oscillator defined previously:

```
def Psi(v, x):
    return N(v)*hermite(v)(x)*np.exp(-0.5*x**2)
```

We now define a function called plot(n) that plots the potential energy as a parabola (black) and spatial probabilities $|\psi_v(x)|^2$ (color) of the normalized quantum Harmonic oscillator wave functions for the integer vibrational quantum number $v = 0$ to $v = n - 1$. Here is the code:

```
def plot(n):
    fig, ax = plt.subplots(figsize=(n,n))

    # Range of x
    xmax = np.sqrt(2*n+1)
    x = np.linspace(-xmax, xmax, 1000)

    for v in range(n):

        # plot potential energy function 0.5*x**2
        ax.plot(x,0.5*x**2,color='black')
```

```
    # plot spatial probabilities psi squared for each energy
level
    ax.plot(x,Psi(v,x)**2 + v + 0.5)
```

```
    # add lines and labels
    ax.axhline(v + 0.5, color='gray', linestyle='-')
    ax.text(xmax, 1.2*(v+0.5),  f"v={v}")
```

```
  ax.set_xlabel('x')
  ax.set_ylabel('$|\psi_v(x)|^2$')
```

We now call the `plot(5)` function. *Figure 6.5* shows the result:

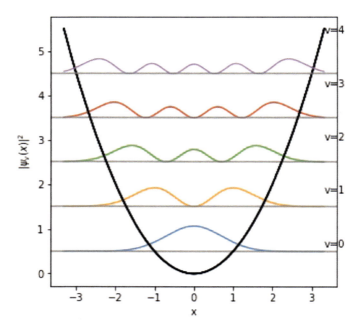

Figure 6.5 – Potential energy (black parabola) and spatial probabilities (color) of the normalized quantum harmonic oscillator wavefunctions for $v = 0$ to $v = 4$

For larger values of the displacement around the equilibrium position, the real potential energy is better approximated by an anharmonic oscillator, and it has the following form:

$$E_{vib}(v) = h\,\omega\left(v + \frac{1}{2}\right) - \chi\,h\,\omega\left(v + \frac{1}{2}\right)^2$$

where χ is the anharmonicity constant. The separations between energy levels decrease with increasing vibrational quantum number v.

The rotational energy at an internuclear distance R can be approximated by the following power series expansion [Demtröder]:

$$E_{rotational} = \frac{J(J+1)\hbar^2}{2MR^2} = \frac{J(J+1)\hbar^2}{2MR_e^2} - \frac{J^2(J+1)^2\hbar^4}{2M^2kR_e^6} + \frac{3J^3(J+1)^3\hbar^6}{2M^3k^2R_e^{10}} + \cdots$$

where k is a constant pertaining to the restoring force that binds the two atoms together.

This expression can be written in terms of rotational frequencies [Demtröder]:

$$F_{rotational}(J) = B_eJ(J+1) - D_eJ^2(J+1)^2 + H_eJ^3(J+1)^3 + \cdots$$

where B_e is the rotational constant, and D_e and H_e are centrifugal constants:

$$B_e = \frac{\hbar}{4\pi cMR_e^2}$$

$$D_e = \frac{\hbar^3}{4\pi ckM^2R_e^6}$$

$$H_e = \frac{3\hbar^5}{4\pi ck^2M^3R_e^{10}}$$

We define the Python Frot() function, which computes the rotational energy levels of the hydrogen molecule using an expression obtained in Ref. [Campargue]:

```
def Frot(J, Be, De, He, show=False):
    F = Be*J*(J+1) + De*J**2*(J+1)**2 - He*J*3*(J+1)**3*10e-5
    if show:
        print("{} {:.2f}".format(J, F))
    return F
```

We set up a dictionary, `rov`, with the vibrational quantum number v as key and the following values:

- Ground state $E(v, J = 0)$ energy computed by Komasa et al. in 2011 [Komasa]
- Rovibrational parameters of the vibrational levels $v = 0$ to $v = 13$ computed by Campargue in 2011 [Campargue]

Here is the code:

```
#        v    E(v,J=0)      Ee          Be        De         He
rms
rov = {0: (36118.0696, 0.0,        59.33289, 0.045498, 4.277,
3.4),
       1: (31956.9034, 4161.1693,  56.37318, -0.043961, 4.168,
3.2),
       2: (28031.0670, 8087.0058,  53.47892, -0.042523, 4.070,
3.2),
       3: (24335.6787, 11782.3940, 50.62885, -0.041175, 3.963,
3.2),
       4: (20867.7039, 15250.3688, 47.79997, -0.039927, 3.846,
3.2),
       5: (17626.1400, 18491.9328, 44.96596, -0.038795, 3.717,
3.2),
       6: (14612.2901, 21505.7826, 42.09566, -0.037808, 3.571,
3.1),
       7: (11830.1543, 24287.9184, 39.15105, -0.037004, 3.399,
3.1),
       8: (9286.9790,  26831.0937, 36.08416, -0.036451, 3.187,
3.1),
       9: (6994.0292,  29124.0436, 32.83233, -0.036251, 2.902,
3.2)}
```

We compute the vibrational energy levels of the hydrogen molecule for the ground rotational state $(J = 0)$ and for the first excited rotational state $(J = 1)$, and the difference between the two levels for each vibrational quantum number $(v = 0, ...,9)$. Here is the code:

```
print("v  E(v,J=0)    E(v,J=1)    BO Diff.")
for v in range(10):
    E0 = rov[v][0] - Frot(0, rov[v][2], rov[v][3], rov[v][4])
    E1 = rov[v][0] - Frot(1, rov[v][2], rov[v][3], rov[v][4])
```

```
    print("{}  {:.4f}  {:.4f}  {:.4f}".format(v, E0, E1, E0 -
E1))
```

The result in *Figure 6.6* is in good agreement with the result by Komasa et al. [Komasa]:

```
v   E(v,J=0)    E(v,J=1)    BO Diff.
0   36118.0696  35999.5961  118.4735
1   31956.9034  31844.3429  112.5605
2   28031.0670  27924.2890  106.7780
3   24335.6787  24234.5952  101.0835
4   20867.7039  20772.2729  95.4310
5   17626.1400  17536.3722  89.7678
6   14612.2901  14528.2586  84.0315
7   11830.1543  11752.0084  78.1459
8   9286.9790   9214.9641   72.0149
9   6994.0292   6928.5165   65.5127
```

Figure 6.6 – Vibrational energy levels of the hydrogen molecule for $v = 0, ..., 9$, $J = 0$ and $J = 1$

Computing all vibrational-rotational levels of a molecule

We present an outline of the method generally used for computing all vibrational-rotational levels of a molecule [Gaussian_1] [Neese] on a classical computer.

Optimizing the geometry of the molecule

The geometry of the molecule used for vibrational analysis must first be optimized so that the atoms are in equilibrium and have no momentum, that is, all first derivatives of the energy with respect to nuclear Cartesian coordinates (x_i, y_i, z_i) of the atoms are zero:

$$\frac{\partial E}{\partial x_i} = \frac{\partial E}{\partial y_i} = \frac{\partial E}{\partial z_i} = 0$$

Calculating a force constant Hessian matrix

A force constant Hessian matrix is calculated, which holds the second partial derivatives of the energy $E = E_{rotational} + E_{vibrational}$ with respect to displacement of the n atoms in Cartesian coordinates (x_i, y_i, z_i) , $i = 0, ..., n$, $j = 0, ..., n$, for instance:

$$F_{x_i x_j} = \frac{\partial^2 E}{\partial x_i \partial x_j}, \quad F_{x_i y_j} = \frac{\partial^2 E}{\partial x_i \partial y_j}, \quad F_{x_i z_j} = \frac{\partial^2 E}{\partial x_i \partial z_j} \cdots$$

A force constant Hessian matrix is a $3n$ by $3n$ matrix. The second derivatives are calculated as finite differences of the gradient when analytical expressions of these derivatives are not available.

Converting to mass weighted Cartesian coordinates

The following change of Cartesian coordinates is then applied:

$$x_i \rightarrow \sqrt{m_i}\, x_i \,, y_i \rightarrow \sqrt{m_i}\, y_i \,, z_i \rightarrow \sqrt{m_i}\, z_i$$

by dividing by $\sqrt{m_i m_j}$ each element pertaining to atoms i and j in the force constant Hessian matrix.

Diagonalizing the mass weighted Hessian matrix

The mass weighted Hessian matrix is then diagonalized, yielding a set of $3n$ eigenvectors and $3n$ eigenvalues. The vibrational frequencies are then derived from the eigenvalues, which are forces using the equation that gives the harmonic oscillator frequency:

$$\upsilon = \frac{1}{2\pi} \sqrt{\frac{k}{\mu}}$$

where υ is the frequency in s-1, μ is the reduced mass, and k is the force constant. The frequency f in cm-1 is obtained by the relation $f = \dfrac{\upsilon}{c}$, where c is the speed of light, $c = 2.998 \times 10^{10}$ cm s-1:

$$f = \frac{1}{2\pi c} \sqrt{\frac{k}{\mu}}$$

We now illustrate this method by performing a vibrational frequency analysis of the CO2 molecule with Psi4 [Psi4_0], an open source quantum chemistry package. We refer the reader interested in installing this package to the "Get Started with Psi4" [Psi4_1] documentation.

First, we import Psi4 in our Python notebook:

```
import psi4
```

We redirect the output of the Psi4 calculation to a file:

```
psi4.core.set_output_file('psi_CO2_output.txt', False)
```

We specify the amount of memory for the calculation:

```
psi4.set_memory('500 MB')
```

Then we define the geometry of the CO_2 molecule [Psi4_2]:

```
co2 = psi4.geometry("""
symmetry c1
0 1
C 1.6830180 -0.4403696 3.1117942
O 0.5425545 -0.2216001 2.9779653
O 2.8186228 -0.6587208 3.2810031
units angstrom
""")
```

We optimize the geometry of the molecule:

```
psi4.set_options({'reference': 'rhf'})
psi4.optimize('scf/cc-pvdz', molecule=co2)
```

Here is the result, the energy of the ground electronic state of a CO_2 molecule:

```
Optimizer: Optimization complete!
-187.65250930298149
```

We now perform a vibrational frequency analysis:

```
scf_e, scf_wfn = psi4.frequency('scf/cc-pvdz', molecule=co2,
return_wfn=True)
```

We print the frequencies with the following code:

```
for i in range(4):
    print(scf_wfn.frequencies().get(0,i))
```

Here is the result, a list of harmonic vibrational frequencies in cm^{-1}:

761.4181081677268

761.4181227549785

1513.1081106509557

2579.8280005025586

These results are in excellent agreement with those presented by the Molecular Sciences Software Institute in their lesson to build a CO2 molecule [MolSSI]. ChemTube3D provides an interactive 3D animation of the vibrations of carbon dioxide with these vibrations [ChemTube3D].

Figure 6.7 shows an extract of the vibrational frequency analysis performed with Psi4.

```
Vibration                        6                    7                    8
Freq [cm^-1]                761.4181             761.4181            1513.1081
Irrep                                                A                    A
Reduced mass [u]             12.8774              12.8774              15.9949
Force const [mDyne/A]         4.3987               4.3987              21.5761
Turning point v=0 [a0]        0.1108               0.1108               0.0705
RMS dev v=0 [a0 u^1/2]        0.2812               0.2812               0.1995
IR activ [km/mol]            66.0405              66.0405               0.0000
Char temp [K]              1095.5111            1095.5112            2177.0257
  ---------------------------------------------------------------------------
     1   C         -0.00  0.00  0.88    0.88 -0.11  0.00    0.00 -0.00  0.00
     2   O          0.00 -0.00 -0.33   -0.33  0.04 -0.00    0.09  0.70 -0.00
     3   O          0.00 -0.00 -0.33   -0.33  0.04 -0.00   -0.09 -0.70 -0.00

Vibration                        9
Freq [cm^-1]               2579.8280
Irrep                            A
Reduced mass [u]             12.8774
Force const [mDyne/A]        50.4962
Turning point v=0 [a0]        0.0602
RMS dev v=0 [a0 u^1/2]        0.1528
IR activ [km/mol]          1038.4320
Char temp [K]              3711.7981
  ---------------------------------------------------------------------------
     1   C          0.11  0.88 -0.00
     2   O         -0.04 -0.33  0.00
     3   O         -0.04 -0.33 -0.00
```

Figure 6.7 – Vibrational frequency analysis of the CO2 molecule with Psi4 (extract)

We now present the vibrational spectra for ortho-para isomerization of hydrogen molecules calculated with BO and non-BO methods and a comparison with experimental data.

6.3. Vibrational spectra for ortho-para isomerization of hydrogen molecules

Figure 6.8 shows a table comparing the energy spacing in wavenumber pertaining to the vibrational spectra for ortho-para isomerization of hydrogen molecules from the following sources described by columns A to C:

- *A* is the energy in wave numbers produced by non-BO method with 10,000 basis functions [Sharkey].

- *B* is the energy in wave numbers produce by BO method [Komasa].

- C is experimental data in wave numbers [Dabrowski].

For each vibrational level ($v = 0, \dots, 14$) with the differences presented in percent [Sharkey_1] described by D to G:

- $D = (C - A) * 100$
- $E = (C - B) * 100$
- $F = (A - B) * 100$
- $G = (F/B) * 100$

Here is the table [Sharkey_1]:

Comparison of non-BO versus BO Computations to Experiment							
Vibrational quantum number (v)	A	B	C	D	E	F	G
0	118.4925	118.4869	118.5	0.755	1.310	0.5553	0.47%
1	112.5803	112.5744	112.61	2.969	3.560	0.5911	0.53%
2	106.7849	106.7917	106.88	9.514	8.830	0.6836	0.64%
3	101.1034	101.0969	101.15	4.663	5.310	0.6475	0.64%
4	95.4510	95.4441	95.5	4.896	5.590	0.6936	0.73%
5	89.7877	89.7807	89.79	0.225	0.930	0.7048	0.78%
6	84.0515	84.044	84.04	-1.149	-0.400	0.7486	0.89%
7	78.1658	78.1579	78.23	6.422	7.210	0.7882	1.01%
8	72.0347	72.0264	72.02	-1.467	-0.640	0.8271	1.15%
9	65.5322	65.5235	65.55	1.781	2.650	0.8693	1.33%
10	58.4872	58.479	58.46	-2.723	-1.900	0.8235	1.41%
11	50.6596	50.651	50.67	1.038	1.900	0.8624	1.70%
12	41.6832	41.6742	41.73	4.679	5.580	0.9012	2.16%
13	30.9466	30.9374	30.95	0.343	1.260	0.9175	2.96%
14	17.1700	17.1607	17.14	-3.003	-2.070	0.9334	5.44%

Figure 6.8 – Table comparing non-BO versus BO Computations to Experiment [Sharkey_1]

The non-BO method is highly accurate in predicting all states, including excited states, as shown in *Figure 6.9*, which presents columns A and B as a function of the vibrational quantum number. It is consistently lower in the predictions of the energy levels, with the exception of the $v = 2$ excited state, an anomaly attributed to the inaccuracy of the experimental data. There is a 5% breakdown of the BO prediction in the highest bound state ($v = 14$)[Sharkey_1].

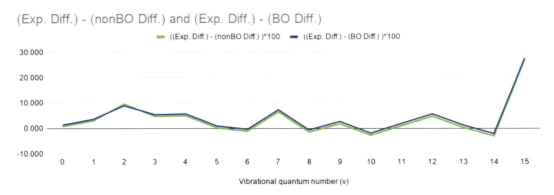

Figure 6.9 – Diagram comparing non-BO versus BO Computations to Experiment [Sharkey_1]

Figure 6.10 shows column G, a plot of the difference between the BO and non-BO computations as a function of the vibrational quantum number (v).

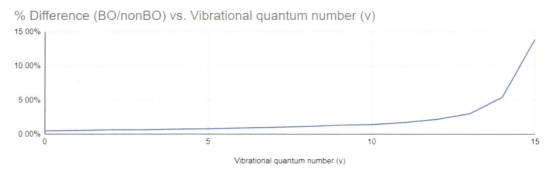

Figure 6.10 – Diagram showing the % difference between BO/non-BO versus vibrational quantum number (v) [Sharkey_1]

Summary

In this chapter, we have recalled the non-BO molecular Hamiltonian and given an outline of a method for extending the very accurate non-BO calculations with ECGs to states where the diatomic molecule is excited to the first rotational state and also vibrationally excited to an arbitrary level. We have shown a vibrational frequency analysis calculation with Psi4 of the carbon dioxide molecule. We have presented the vibrational spectra for ortho-para isomerization of hydrogen molecules calculated with a non-BO method [Sharkey], with a BO method [Komasa], and their comparison with experimental data. The non-BO method is highly accurate in predicting all states, including excited states of the hydrogen molecule. The scaling of the wave functions used for both BO and non-BO methods has a factorial dependence and is considered an NP-hard problem.

Questions

1. True or False: The computational complexity of determining molecular structure and vibrational spectra increases exponentially as a function of the number of atoms.

2. What does the acronym ECG stand for?

3. True or False: In the harmonic oscillator approximation, the force required to extend the spring is proportional to the extension.

4. True or False: When the vibrational energy is high, the harmonic oscillator approximation is no longer valid.

5. True or False: The geometry of the molecule used for vibrational analysis must first be optimized so that the atoms are in equilibrium and have no momentum.

Answers

1. True

2. Explicitly correlated Gaussian function

3. True

4. True

5. True

References

[Adamowicz_1] Sergiy Bubin, Michele Pavanello, Wei-Cheng Tung, Keeper L. Sharkey, and Ludwik Adamowicz, Born–Oppenheimer and Non-Born–Oppenheimer, Atomic and Molecular Calculations with Explicitly Correlated Gaussians, Chem. Rev. 2013, 113, 1, 36–79, October 1, 2012, `https://doi.org/10.1021/cr200419d`

[Adamowicz_2] Sergiy Bubin and Ludwik Adamowicz, Computer program ATOM-MOL-nonBO for performing calculations of ground and excited states of atoms and molecules without assuming the Born–Oppenheimer approximation using all-particle complex explicitly correlated Gaussian functions, J. Chem. Phys. 152, 204102 (2020), `https://doi.org/10.1063/1.5144268`

[Adamowicz_4] Jim Mitroy, Sergiy Bubin, Wataru Horiuchi, Yasuyuki Suzuki, Ludwik Adamowicz, Wojciech Cencek, Krzysztof Szalewicz, Jacek Komasa, D. Blume, and Kálmán Varga, Rev. Mod. Phys. 85, 693 – Published 6 May 2013, Theory and application of explicitly correlated Gaussians, `https://journals.aps.org/rmp/abstract/10.1103/RevModPhys.85.693`

[Campargue] Alain Campargue, Samir Kassi, Krzysztof Pachucki and Jacek Komasa, The absorption spectrum of H2: CRDS measurements of the (2-0) band, review of the literature data and accurate ab initio line list up to 35000 cm-1, Physical Chemistry Chemical Physics, 13 Sep 2011, Table 5. Rovibrational parameters of the V=0-13 vibrational levels of H2 obtained from the fit of the J=0-7 energy levels calculated in Ref. [36], `https://www.fuw.edu.pl/~krp/papers/camparge.pdf`

[ChemTube3D] Vibrations of Carbon Dioxide, `https://www.chemtube3d.com/vibrationsco2`

[D4.1 VA Beta] Arseny Kovyrshin, AstraZeneca AB R&D, Giorgio Silvi, HQS Quantum Simulations GmbH, D4.1: VA Beta and BBO Beta, NExt ApplicationS of Quantum Computing, 23 Nov 2021, `https://www.neasqc.eu/wp-content/uploads/2022/01/NEASQC_D4.1_VA-Beta-and-BBO-Beta-R1.0.pdf`, `https://github.com/NEASQC/Variationals_algorithms`

[Dabrowski] Dabrowski, The Lyman and Werner Bands of H2, Can. J. Phys. 62, 1639 (1984) Table 5. Observed energy levels of the $X^1 \Sigma_g^+$ of H_2, `https://doi.org/10.1139/p84-210`

[Demtröder] Atoms, Molecules and Photons, Wolfgang Demtröder, Second Edition, Springer, 9.5. Rotation and Vibration of Diatomic Molecules, Springer, ISBN-13: 978-3642102974

[Gaussian_1] Joseph W. Ochterski, Gaussian, Vibrational Analysis in Gaussian, `https://gaussian.com/vib/`

[Hartke] Hartke, T., Oreg, B., Jia, N. et al. Quantum register of fermion pairs. Nature 601, 537–541 (2022). `https://doi.org/10.1038/s41586-021-04205-8`

[Komasa] Komasa et al., Quantum Electrodynamics Effects in Rovibrational Spectra of Molecular Hydrogen J. Chem. Theory Comput. 2011, 7, 10, 3105–3115, Table 1. Theoretically Predicted Dissociation Energies {in cm^(-1)} of All 302 Bound States of H_2. `https://doi.org/10.1021/ct200438t`

[Maytus] Edit Mátyus, Edit Mátyus (2019) Pre-Born–Oppenheimer molecular structure theory, Molecular Physics, 117:5, 590-609, DOI: 10.1080/00268976.2018.1530461, `https://doi.org/10.1080/00268976.2018.1530461`

[MolSSI] Basis set convergence of molecular properties: Geometry and Vibrational Frequency, Molecular Sciences Software Institute (MolSSI), `http://education.molssi.org/qm-tools/04-vib-freq/index.html`

[Neese] Vibrational Spectroscopy, Frank Neese from the Max Planck Institute for Chemical Energy Conversion, 2014 summer school, `https://www.youtube.com/watch?v=iJjg2L1F8I4`

[Psi4_0] Psi4 manual master index, `https://psicode.org/psi4manual/master/index.html`

[Psi4_1] Get Started with PSI4, `https://psicode.org/installs/v15/`

[Psi4_2] Test case for Binding Energy of C4H5N (Pyrrole) with CO2 using MP2/def2-TZVPP, `https://github.com/psi4/psi4/blob/master/samples/mp2-def2/input.dat`

[Psi4_3] Smith DGA, Burns LA, Simmonett AC, Parrish RM, Schieber MC, Galvelis R, Kraus P, Kruse H, Di Remigio R, Alenaizan A, James AM, Lehtola S, Misiewicz JP, Scheurer M, Shaw RA, Schriber JB, Xie Y, Glick ZL, Sirianni DA, O'Brien JS, Waldrop JM, Kumar A, Hohenstein EG, Pritchard BP, Brooks BR, Schaefer HF 3rd, Sokolov AY, Patkowski K, DePrince AE 3rd, Bozkaya U, King RA, Evangelista FA, Turney JM, Crawford TD, Sherrill CD. Psi4 1.4: Open-source software for high-throughput quantum chemistry. J Chem Phys. 2020 May 14;152(18):184108. doi: 10.1063/5.0006002. PMID: 32414239; PMCID: PMC7228781. `https://www.ncbi.nlm.nih.gov/pmc/articles/PMC7228781/pdf/JCPSA6-000152-184108_1.pdf`

[Sawaya] Nicolas P. D. Sawaya, Francesco Paesani, Daniel P. Tabor, Near- and long-term quantum algorithmic approaches for vibrational spectroscopy, 1 Feb 2021, arXiv:2009.05066 [quant-ph], `https://arxiv.org/abs/2009.05066`

[Schiffer_1] Fabijan Pavošević, Tanner Culpitt, and Sharon Hammes-Schiffer, Multicomponent Quantum Chemistry: Integrating Electronic and Nuclear Quantum Effects via the Nuclear–Electronic Orbital Method, Chem. Rev. 2020, 120, 9, 4222–4253, https://doi.org/10.1021/acs.chemrev.9b00798

[Schiffer_2] Kurt R. Brorsen, Andrew Sirjoosingh, Michael V. Pak, and Sharon Hammes-Schiffer, Nuclear-electronic orbital reduced explicitly correlated Hartree-Fock approach: Restricted basis sets and open-shell systems, J. Chem. Phys. 142, 214108 (2015), https://doi.org/10.1063/1.4921304

[SciPy_0] SciPy, https://scipy.org/

[Sharkey] K. Sharkey et. al. Non-Born-Oppenheimer method for direct variational calculations of diatomic first excited rotational states using explicitly correlated all-particle Gaussian functions Physical Review A, 88, 032513 (2013, Table I. Total energies (in hartrees) of the (v,0) and (v,1) states of H_2. https://journals.aps.org/pra/abstract/10.1103/PhysRevA.88.032513

[Sharkey_1] K. Sharkey, Molecular-hydrogen poster, QLEAN™, https://qlean.world/molecular-hydrogen-poster

[Veis] Libor Veis, Jakub Višňák, Hiroaki Nishizawa, Hiromi Nakai, Jiří Pittner, Quantum chemistry beyond Born–Oppenheimer approximation on a quantum computer: A simulated phase estimation study, International Journal of Quantum Chemistry, 22 June 2016, https://doi.org/10.1002/qua.25176

7
Conclusion

"I have no special talents. I am only passionately curious."

– Albert Einstein

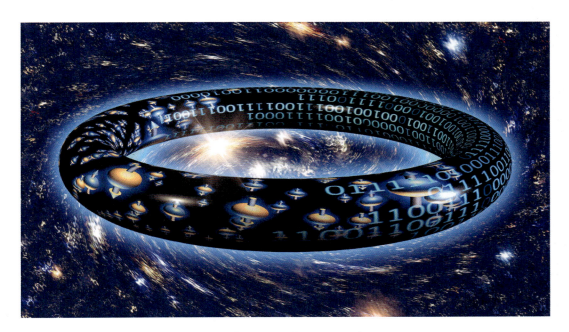

Figure 7.1 – Circular dependency of quantum chemistry and quantum computing [authors]

7.1. Quantum computing

The initial quantum circuit model of computation as a time-ordered sequence of logical quantum gates performing a unitary evolution of the state of a quantum register has evolved into a tightly integrated model of dynamic quantum circuits that allow concurrent classical processing of mid-circuit measurement results [Corcoles] [IBM_mid]. This new paradigm of dynamic quantum computation paves the way to a smooth transition from classical to quantum-boosted computing.

An often-overlooked potential quantum advantage is the energy efficiency of quantum computation [Auffeves] [Q Daily] [Quantum_AI] [Thibault]. The quantum supremacy experiment [Arute] involved an energy consumption three orders of magnitude smaller than the one a high-performance computer would require to achieve an exact computation of the expected result of the experiment. Determining the conditions under which an energetic quantum advantage could be achieved is an open research topic [Auffeves].

The recent demonstration of a new quantum register of fermion pairs where information is stored in the vibrational motion of atom pairs held in a superposition of two vibrational states opens the perspective of programmable quantum simulators of molecules [Hartke]. The accurate calculations of the vibrational spectra of molecules, which have applications from astrochemistry to biochemistry and climate change mitigation, might be easier to achieve on a quantum computer than electronic energy calculations [Sawaya].

7.2. Quantum chemistry

In *Chapter 4, Molecular Hamiltonians and Chapter 5, Variational Quantum Eigensolver (VQE) Algorithm*, we illustrated some of the methods of quantum computational chemistry with Python and open-source quantum chemistry packages PySCF, ASE, PyQMC, and Qiskit, solving for the ground state energy level and plotting the BOPES of the hydrogen molecule, the lithium hydride molecule, and the macro-molecule ONCHHC. With the simplest basis (STO-3G), and with a noise-free simulation of a quantum circuit (statevector simulator), the different methods of calculations were in good agreement.

Scientific or industrial applications require highly accurate relative energy estimates of about 1 milli-Hartree (mHA) or even 0.1 mHA of chemical reaction mechanisms. The same accuracy is required for the total electronic energy of each molecule species involved in a chemical reaction of interest [Burg]. The authors of a benchmark of algorithms for calculating the electronic structures of molecules of relevance to prebiotic chemistry have concluded that "To utilize VQE and achieve near chemical accuracy will be extremely challenging for NISQ processors" [Lolur]. They point out that the main challenge is the large number of Pauli terms resulting from the mapping of fermionic to qubit Hamiltonian, the large number of variational parameters, and the large number of energy evaluations. Furthermore, to get an accurate energy estimate of 1 mHA, the ansatz for the VQE must be close to the true ground state by less than one in a million [Troyer].

In *Chapter 6, Beyond Born-Oppenheimer,* we explained how non-Born-Oppenheimer (non-BO) calculations include the effects that are needed to make better predictions of chemical states at above ground state vibrations. Implementing these non-BO calculations with innovative hybrid classical-quantum algorithms is an open research topic. Ryan Babbush, head of Google's quantum algorithms team, has developed the first quantized quantum simulations algorithms for chemistry. In the introduction of his presentation of these algorithms [Babbush], he stated that there are lots of contexts where non-BO simulations are important, such as low temperatures where hydrogen bondings are involved, or tunneling or couplings between electrons and nuclei, or to compute dynamics, reactive scattering coefficients, or thermo rate constants directly from quantum dynamics. A study of these algorithms [Su] has shown potential advantages to algorithms in second quantization. However, these algorithms require fault-tolerant quantum computers with several thousand logical qubits able to run quantum circuits with a huge number of gates (10^{11} to 10^{12}), way beyond the capabilities of the current or near-term NISQ-era processors.

References

[Arute] Arute, F., Arya, K., Babbush, R. et al., Quantum supremacy using a programmable superconducting processor, Nature 574, 505–510 (2019), https://doi.org/10.1038/s41586-019-1666-5

[Auffeves] Alexia Auffèves, Optimiser la consommation énergétique des calculateurs quantiques : un défi interdisciplinaire, Reflets phys. N°69 (2021) 16-20, 12 July 2021, https://doi.org/10.1051/refdp/202169016

[Babbush] Ryan Babbush, Feb. 24, 2021, The Promise of First Quantized Quantum Simulations of Chemistry, Google AI Quantum, Chemistry's fault tolerant future is first quantized!, `https://www.youtube.com/watch?v=iugrIX616yg`

[Burg] Vera von Burg, Guang Hao Low, Thomas Häner, Damian S. Steiger, Markus Reiher, Martin Roetteler, Matthias Troyer, Quantum computing enhanced computational catalysis, 3 Mar 2021, 10.1103/PhysRevResearch.3.033055, `https://arxiv.org/abs/2007.14460`

[Corcoles] A. D. Córcoles, Maika Takita, Ken Inoue, Scott Lekuch, Zlatko K. Minev, Jerry M. Chow, and Jay M. Gambetta, Exploiting Dynamic Quantum Circuits in a Quantum Algorithm with Superconducting Qubits, Phys. Rev. Lett. 127, 100501, 31 August 2021, `https://journals.aps.org/prl/abstract/10.1103/PhysRevLett.127.100501`

[Hartke] Hartke, T., Oreg, B., Jia, N. et al., Quantum register of fermion pairs, Nature 601, 537–541 (2022), `https://doi.org/10.1038/s41586-021-04205-8`

[IBM_mid] Mid-Circuit Measurements Tutorial, IBM Quantum systems, `https://quantum-computing.ibm.com/lab/docs/iql/manage/systems/midcircuit-measurement/`

[Lolur] Lolur, Phalgun, Magnus Rahm, Marcus Skogh, Laura García-Álvarez and Göran Wendin, Benchmarking the Variational Quantum Eigensolver through Simulation of the Ground State Energy of Prebiotic Molecules on High-Performance Computers, arXiv:2010.13578v2 [quant-ph], 5 Jan 2021, `https://arxiv.org/pdf/2010.13578.pdf`

[Q Daily] Quantum Technology | Our Sustainable Future, The Quantum Daily, Jul 29, 2021, `https://www.youtube.com/watch?v=iB2_ibvEcsE`

[Quantum_AI] The Quantum AI Sustainability Symposium, Q4Climate, Speakers: Dr. Karl Thibault, Mr. Michał Stęchły, Sep 1, 2021, `https://quantum.ieee.org/conferences/quantum-ai`

[Sawaya] Nicolas P. D. Sawaya, Francesco Paesani, Daniel P. Tabor, Near- and long-term quantum algorithmic approaches for vibrational spectroscopy, 1 Feb 2021, arXiv:2009.05066 [quant-ph], `https://arxiv.org/abs/2009.05066`

[Su] Yuan Su, Dominic W. Berry, Nathan Wiebe, Nicholas Rubin, and Ryan Babbush, Fault-Tolerant Quantum Simulations of Chemistry in First Quantization, 11 Oct. 2021, PRX Quantum 2, 040332, DOI:10.1103/PRXQuantum.2.040332, `https://doi.org/10.1103/PRXQuantum.2.040332`

[Thibault] Casey Berger, Agustin Di Paolo, Tracey Forrest, Stuart Hadfield, Nicolas Sawaya, Michał Stęchły, Karl Thibault, Quantum technologies for climate change: Preliminary assessment, IV. ENERGY EFFICIENCY OF QUANTUM COMPUTERS By Karl Thibault, arXiv:2107.05362 [quant-ph], 23 Jun 2021, https://arxiv.org/abs/2107.05362

[Troyer] Matthias Troyer, Matthias Troyer: Achieving Practical Quantum Advantage in Chemistry Simulations, QuCQC 2021, https://www.youtube.com/watch?v=2MsfbPlKgyI

8
References

[ASE_0] Atomic Simulation Environment (ASE), `https://wiki.fysik.dtu.dk/ase/index.html`

[ASE_1] ASE, The Atoms object, `https://wiki.fysik.dtu.dk/ase/atoms.html`

[ASE_2] ASE Visualization, `https://wiki.fysik.dtu.dk/ase/visualize/visualize.html#module-ase.visualize`

[Aaronson_1] Scott Aaronson, The Limits of Quantum Computers, Scientific American, March 2008, `https://www.scientificamerican.com/article/the-limits-of-quantum-computers/`

[Aaronson_2] Scott Aaronson, The Limits of Quantum Computers (DRAFT), `https://www.scottaaronson.com/writings/limitsqc-draft.pdf`

[Adamowicz_1] Sergiy Bubin, Michele Pavanello, Wei-Cheng Tung, Keeper L. Sharkey, and Ludwik Adamowicz, Born–Oppenheimer and Non-Born–Oppenheimer, Atomic and Molecular Calculations with Explicitly Correlated Gaussians, Chem. Rev. 2013, 113, 1, 36–79, October 1, 2012, `https://doi.org/10.1021/cr200419d`

[Adamowicz_2] Sergiy Bubin and Ludwik Adamowicz, Computer program ATOM-MOL-nonBO for performing calculations of ground and excited states of atoms and molecules without assuming the Born–Oppenheimer approximation using all-particle complex explicitly correlated Gaussian functions, J. Chem. Phys. 152, 204102 (2020), `https://doi.org/10.1063/1.5144268`

[Adamowicz_3] Tung WC, Pavanello M, Adamowicz L., Very accurate potential energy curve of the LiH molecule. TABLE I. Comparison of the convergence of the BO energy, in Eh, for the ground state of LiH molecule at R = 3.015 bohrs, J Chem Phys. 2011 Feb 14;134(6):064117. doi: 10.1063/1.3554211, https://doi.org/10.1063/1.3554211

[Adamowicz_4] Jim Mitroy, Sergiy Bubin, Wataru Horiuchi, Yasuyuki Suzuki, Ludwik Adamowicz, Wojciech Cencek, Krzysztof Szalewicz, Jacek Komasa, D. Blume, and Kálmán Varga, Rev. Mod. Phys. 85, 693 — Published 6 May 2013, Theory and application of explicitly correlated Gaussians, https://journals.aps.org/rmp/abstract/10.1103/RevModPhys.85.693

[Africa21] IBM Quantum Challenge Africa 2021, https://github.com/qiskit-community/ibm-quantum-challenge-africa-2021

[Arute] Arute, F., Arya, K., Babbush, R. et al., Quantum supremacy using a programmable superconducting processor, Nature 574, 505–510 (2019), https://doi.org/10.1038/s41586-019-1666-5

[Auffeves] Alexia Auffèves, Optimiser la consommation énergétique des calculateurs quantiques : un défi interdisciplinaire, Reflets phys. N°69 (2021) 16-20, 12 July 2021, https://doi.org/10.1051/refdp/202169016

[Babbush] Ryan Babbush, Feb. 24, 2021, The Promise of First Quantized Quantum Simulations of Chemistry, Google AI Quantum, Chemistry's fault tolerant future is first quantized!, https://www.youtube.com/watch?v=iugrIX616yg

[Balmer_series] Balmer Series, Wikipedia, https://en.wikipedia.org/wiki/Balmer_series

[Bell_1] Bell, J. S., On the Einstein Podolsky Rosen Paradox, Physics Physique Fizika 1, 195: 195–200, 1964, https://doi.org/10.1103/PhysicsPhysiqueFizika.1.195

[Bell_2] "Chapter 2: On the Einstein-Podolsky-Rosen paradox". Speakable and Unspeakable in Quantum Mechanics: Collected Papers on Quantum Philosophy (Alain Aspect introduction to 1987 ed.), Reprinted in JS Bell (2004), Cambridge University Press. pp. 14–21. ISBN 978-0521523387

[Benioff] Benioff, P., The computer as a physical system: A microscopic quantum mechanical Hamiltonian model of computers as represented by Turing machines, https://doi.org/10.1007/BF01011339

[Bittel] Lennart Bittel and Martin Kliesch, Training variational quantum algorithms is NP-hard — even for logarithmically many qubits and free fermionic systems, DOI:10.1103/PhysRevLett.127.120502, 18 Jan 2021, https://doi.org/10.1103/PhysRevLett.127.120502

[Bohr_1] N. Bohr, I., On the Constitution of Atoms and Molecules, Philosophical Magazine, 26, 1-25 (July 1913), DOI: 10.1080/14786441308634955

[Bohr_2] Bohr's shell model, Britannica, https://www.britannica.com/science/atom/Bohrs-shell-model#ref496660

[Born_1] Born, M., Jordan, P. Zur Quantenmechanik, Z. Physik 34, 858–888 (1925), https://doi.org/10.1007/BF01328531

[Bravyi] Sergey Bravyi, Jay M. Gambetta, Antonio Mezzacapo, Kristan Temme, Tapering off qubits to simulate fermionic Hamiltonians, arXiv:1701.08213v1, 27 Jan 2017, https://arxiv.org/pdf/1701.08213.pdf

[Bubin] Bubin, S., Cafiero, M., & Adamowicz, L., Non-Born-Oppenheimer variational calculations of atoms and molecules with explicitly correlated Gaussian basis functions, Advances in Chemical Physics, 131, 377-475, https://doi.org/10.1002/0471739464.ch6

[Burg] Vera von Burg, Guang Hao Low, Thomas Häner, Damian S. Steiger, Markus Reiher, Martin Roetteler, Matthias Troyer, Quantum computing enhanced computational catalysis, 3 Mar 2021, 10.1103/PhysRevResearch.3.033055, https://arxiv.org/abs/2007.14460

[Bushmanov] Sergey Bushmanov, How to plot a histogram using Matplotlib in Python with a list of data?, Stack Overflow, https://stackoverflow.com/questions/33203645/how-to-plot-a-histogram-using-matplotlib-in-python-with-a-list-of-data

[Byjus] BYJU'S, Hydrogen Spectrum, Wavelength, Diagram, Hydrogen Emission Spectrum, https://byjus.com/chemistry/hydrogen-spectrum/#

[CERN_quark] CERN, Voyage into the world of atoms, https://www.youtube.com/watch?v=7WhRJV_bAiE

[Campargue] Alain Campargue, Samir Kassi, Krzysztof Pachucki and Jacek Komasa, The absorption spectrum of H2: CRDS measurements of the (2-0) band, review of the literature data and accurate ab initio line list up to 35000 cm-1, Table 5. Rovibrational parameters of the V=0-13 vibrational levels of H2 obtained from the fit of the J=0-7 energy levels calculated in Ref. [36], Physical Chemistry Chemical Physics, 13 Sep 2011, https://www.fuw.edu.pl/~krp/papers/camparge.pdf

[Cao] Yudong Cao, Jonathan Romero, Jonathan P. Olson, Matthias Degroote, Peter D. Johnson, Mária Kieferová, Ian D. Kivlichan, Tim Menke, Borja Peropadre, Nicolas P. D. Sawaya, Sukin Sim, Libor Veis, Alán Aspuru-Guzik, Quantum Chemistry in the Age of Quantum Computing, Chem. Rev. 2019, 119, 19, 10856–10915, Aug 30, 2019, `https://doi.org/10.1021/acs.chemrev.8b00803`

[Chem-periodic] Chemistry LibreTexts, 5.17: Electron Configurations and the Periodic Table, `https://chem.libretexts.org/Bookshelves/General_Chemistry/Book%3A_ChemPRIME_(Moore_et_al.)/05%3A_The_Electronic_Structure_of_Atoms/5.17%3A_Electron_Configurations_and_the_Periodic_Table`

[ChemChiral] 5.1 Chiral Molecules, Chemistry LibreTexts, 5 Jul 2015, `https://chem.libretexts.org/Bookshelves/Organic_Chemistry/Map%3A_Organic_Chemistry_(Vollhardt_and_Schore)/05._Stereoisomers/5.1%3A_Chiral__Molecules`

[ChemTube3D] Vibrations of Carbon Dioxide, `https://www.chemtube3d.com/vibrationsco2`

[Chem_spectr] Chemistry LibreTexts, 7.3: The Atomic Spectrum of Hydrogen, `https://chem.libretexts.org/Courses/Solano_Community_College/Chem_160/Chapter_07%3A_Atomic_Structure_and_Periodicity/7.03_The_Atomic_Spectrum_of_Hydrogen`

[Chen] Sija Chen, Quantum Monte Carlo Methods, Maplesoft, `https://fr.maplesoft.com/Applications/Detail.aspx?id=154748`

[Chiew] Mitchell Chiew and Sergii Strelchuk, Optimal fermion-qubit mappings, arXiv:2110.12792v1 [quant-ph], 25 Oct 2021, `https://arxiv.org/pdf/2110.12792.pdf`

[Clay] Millenium problems, `https://www.claymath.org/millennium-problems`

[Cmap] Choosing Colormaps in Matplotlib, `https://matplotlib.org/stable/tutorials/colors/colormaps.html`

[Comp_Zoo] Complexity Zoo, `https://complexityzoo.net/Complexity_Zoo`

[Corcoles] A. D. Córcoles, Maika Takita, Ken Inoue, Scott Lekuch, Zlatko K. Minev, Jerry M. Chow, and Jay M. Gambetta, Exploiting Dynamic Quantum Circuits in a Quantum Algorithm with Superconducting Qubits, Phys. Rev. Lett. 127, 100501, 31 August 2021, `https://journals.aps.org/prl/abstract/10.1103/PhysRevLett.127.100501`

[Crockett] Christopher Crockett, Superpositions of Chiral Molecules, September 14, 2021, Physics 14, s108, https://physics.aps.org/articles/v14/s108

[D4.1 VA Beta] Arseny Kovyrshin, AstraZeneca AB R&D, Giorgio Silvi, HQS Quantum Simulations GmbH, D4.1: VA Beta and BBO Beta, NExt ApplicationS of Quantum Computing, 23 Nov 2021, https://www.neasqc.eu/wp-content/uploads/2022/01/NEASQC_D4.1_VA-Beta-and-BBO-Beta-R1.0.pdf , https://github.com/NEASQC/Variationals_algorithms

[Dabrowski] Dabrowski, The Lyman and Werner Bands of H2, Table 5. Observed energy levels of the $X^1 \Sigma_g^+$ of H_2, Can. J. Phys. 62, 1639 (1984), https://doi.org/10.1139/p84-210

[Dagrada] Mario Dagrada, Improved quantum Monte Carlo simulations : from open to extended systems, Materials Science [cond-mat.mtrl-sci]. Université Pierre et Marie Curie - Paris VI; Universidad Nacional de San Martín, 2016. English. ⟨NNT : 2016PA066349⟩. ⟨tel-01478313⟩, https://tel.archives-ouvertes.fr/tel-01478313/document

[Daskalatis] Costis Daskalakis, Equilibrium Computation & the Foundations of Deep Learning, Costis Daskalakis on Foundation of Data Science Series, Feb 18, 2021, https://www.youtube.com/watch?v=pDangP47ftE

[De Keijzer] de Keijzer, R. J. P. T., Colussi, V. E., Škorić, B., & Kokkelmans, S. J. J. M. F. (2021), Optimization of the Variational Quantum Eigensolver for Quantum Chemistry Applications, arXiv, 2021, [2102.01781], https://arxiv.org/abs/2102.01781

[Demtröder] Atoms, Molecules and Photons, 9.5. Rotation and Vibration of Diatomic Molecules, Wolfgang Demtröder, Second Edition, Springer, Springer, ISBN-13: 978-3642102974

[Deutsch-Jozsa] David Deutsch and Richard Jozsa, Rapid solutions of problems by quantum computation, Proceedings of the Royal Society of London A. 439: 553 558, https://doi.org/10.1098/rspa.1992.0167

[DiVincenzo] David P. DiVincenzo, The Physical Implementation of Quantum Computation, 10.1002/1521-3978(200009)48:9/11<771::AID-PROP771>3.0.CO;2-E, https://arxiv.org/abs/quant-ph/0002077

[Dirac_2] Dirac, P.A.M., The physical interpretation of the quantum dynamics, Proc. R. Soc. Lond. A 1927, 113, 621–641, https://doi.org/10.1098/rspa.1927.0012

[Dowling] Jonathan P. Dowling and Gerard J. Milburn, Quantum technology: the second quantum revolution, Royal Society, 20 June 2003, https://doi.org/10.1098/rsta.2003.1227

[E_mass] Fundamental physical constants, electron mass, NIST, https://physics.nist.gov/cgi-bin/cuu/Value?me|search_for=electron+mass

[Ebomwonyi] Ebomwonyi, Osarodion, A Quantum Monte Carlo Calculation of the Ground State Energy for the Hydrogen Molecule Using the CASINO Code, 2013, Table 3.1: Comparative analysis of the ground state energies for the hydrogen molecule by different researchers, https://www.semanticscholar.org/paper/A-Quantum-Monte-Carlo-Calculation-of-the-Ground-for-Ebomwonyi/5316eb86f39cf4fa0a8fd06d136aac4db1105ad4

[Fearnley] John Fearnley (University of Liverpool), Paul W. Goldberg (University of Oxford), Alexandros Hollender (University of Oxford), and Rahul Savani (University of Liverpool), The Complexity of Gradient Descent: CLS = PPAD ∩ PLS, STOC 2021: Proceedings of the 53rd Annual ACM SIGACT Symposium on Theory of Computing, June 2021 Pages 46–59, https://doi.org/10.1145/3406325.3451052

[Freeman] Freedman–Diaconis rule, Wikipedia, https://en.wikipedia.org/wiki/Freedman%E2%80%93Diaconis_rule

[Gard] Gard, B.T., Zhu, L., Barron, G.S. et al., Efficient symmetry-preserving state preparation circuits for the variational quantum eigensolver algorithm, npj Quantum Inf 6, 10 (2020), https://doi.org/10.1038/s41534-019-0240-1

[Gaussian_1] Joseph W. Ochterski, Gaussian, Vibrational Analysis in Gaussian, https://gaussian.com/vib/

[Getty] Girl looking up, https://media.gettyimages.com/photos/you-learn-something-new-every-day-picture-id523149221?k=20&m=523149221&s=612x612&w=0&h=7ZFg6ETuKlqr1nzi98IBNz-uYXccQwiuNKEk0hGKKIU=

[Gorelov] Vitaly Gorelov, Quantum Monte Carlo methods for electronic structure calculations: application to hydrogen at extreme conditions, 1.4.1 Variational Monte Carlo (VMC), https://tel.archives-ouvertes.fr/tel-03045954/document

[Grok] Grok the Bloch Sphere, https://javafxpert.github.io/grok-bloch/

[H5py] Quick Start Guide, https://docs.h5py.org/en/stable/quick.html

[Hartke] Hartke, T., Oreg, B., Jia, N. et al., Quantum register of fermion pairs, Nature 601, 537–541 (2022), https://doi.org/10.1038/s41586-021-04205-8

[Hill] Learning Scientific Programming with Python, Chapter 2: The Core Python Language I, Problems, P2.5, Electronic configurations, https://scipython.com/book/chapter-2-the-core-python-language-i/questions/problems/p25/electronic-configurations/

[IBM_CEO] IBM CEO: Quantum computing will take off 'like a rocket ship' this decade, Fast Company, Sept 28, 2021., https://www.fastcompany.com/90680174/ibm-ceo-quantum-computing-will-take-off-like-a-rocket-ship-this-decade

[IBM_comp1] Welcome to IBM Quantum Composer, https://quantum-computing.ibm.com/composer/docs/iqx/

[IBM_comp2] IBM Quantum Composer, https://quantum-computing.ibm.com/composer/files/new

[IBM_mid] Mid-Circuit Measurements Tutorial, IBM Quantum systems, https://quantum-computing.ibm.com/lab/docs/iql/manage/systems/midcircuit-measurement/

[Intro_BOA_1] M. Born, J.R. Oppenheimer, On the Quantum Theory of Molecules, https://www.theochem.ru.nl/files/dbase/born-oppenheimer-translated-s-m-blinder.pdf

[Intro_BOA_2] M. Born and R. J. Oppenheimer, Zur Quantentheorie der Molekeln, Annalen der physik, 20, 457-484 (August 1927), https://doi.org/10.1002/andp.19273892002

[Kaplan] Ilya G. Kaplan, Modern State of the Pauli Exclusion Principle and the Problems of Its Theoretical Foundation, Symmetry 2021, 13(1), 21, https://doi.org/10.3390/sym13010021

[Knill] Emanuel Knill, Raymond Laflamme, A Theory of Quantum Error-Correcting Codes, https://arxiv.org/abs/quant-ph/9604034

[Komasa] Komasa et al., Quantum Electrodynamics Effects in Rovibrational Spectra of Molecular Hydrogen J. Chem. Theory Comput. 2011, 7, 10, 3105–3115, Table 1. Theoretically Predicted Dissociation Energies {in cm^(-1)} of All 302 Bound States of H_2, https://doi.org/10.1021/ct200438t, https://www.fuw.edu.pl/~krp/papers/H2D2v18.pdf

[Lolur] Lolur, Phalgun, Magnus Rahm, Marcus Skogh, Laura García-Álvarez and Göran Wendin, Benchmarking the Variational Quantum Eigensolver through Simulation of the Ground State Energy of Prebiotic Molecules on High-Performance Computers, arXiv:2010.13578v2 [quant-ph], 5 Jan 2021, https://arxiv.org/pdf/2010.13578.pdf

[Lucr_1] Lucretius on the Nature of Things, Literally translated into English prose by the Rev. John Selby Watson, M.A., London 1870, `https://www.google.fr/books/edition/Lucretius_On_the_Nature_of_Things/59HTAAAAMAAJ?hl=en&gbpv=1&printsec=frontcover`

[Lucr_2] Thomas Nail, Lucretius: Our Contemporary, 15 Feb 2019, `https://www.youtube.com/watch?v=VMrTk1A2GX8`

[Lucr_3] David Goodhew, Lucretius lecture, Life, love, death and atomic physics, `https://www.youtube.com/watch?v=mJZZd3f_-oE`

[Lyman_series] Lyman series, From Wikipedia, `https://en.wikipedia.org/wiki/Lyman_series`

[MIT_QC_1981] MIT Endicott House, The Physics of Computation Conference, Image "Physics of Computation Conference, Endicott House MIT May 6-8, 1981", Mar 21, 2018, `https://mitendicotthouse.org/physics-computation-conference/`

[Maytus] Edit Mátyus, Edit Mátyus (2019) Pre-Born — Oppenheimer molecular structure theory, Molecular Physics, 117:5, 590-609, DOI: 10.1080/00268976.2018.1530461, `https://doi.org/10.1080/00268976.2018.1530461`

[Mezzacapo] Antonio Mezzacapo, Simulating Chemistry on a Quantum Computer, Part I, Qiskit Global Summer School 2020, IBM Quantum, Qiskit, Introduction to Quantum Computing and Quantum Hardware, `https://qiskit.org/learn/intro-qc-qh/`, Lecture Notes 8, `https://github.com/qiskit-community/intro-to-quantum-computing-and-quantum-hardware/blob/master/lectures/introqcqh-lecture-notes-8.pdf?raw=true`

[Micr_Algebra] Linear algebra, QuantumKatas/tutorials/LinearAlgebra/, `https://github.com/microsoft/QuantumKatas/tree/main/tutorials/LinearAlgebra`

[Micr_Complex] Complex arithmetic, QuantumKatas/tutorials/ComplexArithmetic/, `https://github.com/microsoft/QuantumKatas/tree/main/tutorials/ComplexArithmetic`

[MolSSI] Basis set convergence of molecular properties: Geometry and Vibrational Frequency, Molecular Sciences Software Institute (MolSSI), `http://education.molssi.org/qm-tools/04-vib-freq/index.html`

[Neese] Vibrational Spectroscopy, Frank Neese from the Max Planck Institute for Chemical Energy Conversion, 2014 summer school, `https://www.youtube.com/watch?v=iJjg2L1F8I4`

[NumPy] NumPy: the absolute basics for beginners, `https://numpy.org/doc/stable/user/absolute_beginners.html`

[Neutron-electron-mass-ratio] neutron-electron mass ratio, NIST, `https://physics.nist.gov/cgi-bin/cuu/Value?mnsme`

[Orb_Approx] Definition of Orbital Approximation, `https://www.chemicool.com/definition/orbital-approximation.html`

[Panagiotis] Panagiotis Kl. Barkoutsos, Jerome F. Gonthier, Igor Sokolov, Nikolaj Moll, Gian Salis, Andreas Fuhrer, Marc Ganzhorn, Daniel J. Egger, Matthias Troyer, Antonio Mezzacapo, Stefan Filipp, Ivano Tavernelli, Quantum algorithms for electronic structure calculations: Particle-hole Hamiltonian and optimized wave-function expansions, Phys. Rev. A 98, 022322 – Published 20 August 2018, DOI: 10.1103/PhysRevA.98.022322, `https://link.aps.org/doi/10.1103/PhysRevA.98.022322`

[Part_1] List of particles, Wikipedia, `https://en.wikipedia.org/wiki/List_of_particles`

[Pauling] L. Pauling and E. B. Wilson, Introduction to Quantum Mechanics with Applications to Chemistry, Dover (1935)

[Pease] Christopher Pease, An Overview of Monte Carlo Methods, Towards Data Science, `https://towardsdatascience.com/an-overview-of-monte-carlo-methods-675384eb1694`

[Phys5250] Addition of angular momentum, University of Colorado, PHYS5250, `https://physicscourses.colorado.edu/phys5250/phys5250_fa19/lecture/lec32-addition-angular-momentum/`

[PoorLeno] File:Hydrogen Density Plots.png, from Wikipedia, `https://en.wikipedia.org/wiki/File:Hydrogen_Density_Plots.png`

[Preskill_40y] John Preskill, Quantum computing 40 years later, `https://arxiv.org/abs/2106.10522`

[Psi4_0] Psi4 manual master index, `https://psicode.org/psi4manual/master/index.html`

[Psi4_1] Get Started with PSI4, `https://psicode.org/installs/v15/`

[Psi4_2] Test case for Binding Energy of C4H5N (Pyrrole) with CO2 using MP2/def2-TZVPP, `https://github.com/psi4/psi4/blob/master/samples/mp2-def2/input.dat`

[Psi4_3] Smith DGA, Burns LA, Simmonett AC, Parrish RM, Schieber MC, Galvelis R, Kraus P, Kruse H, Di Remigio R, Alenaizan A, James AM, Lehtola S, Misiewicz JP, Scheurer M, Shaw RA, Schriber JB, Xie Y, Glick ZL, Sirianni DA, O'Brien JS, Waldrop JM, Kumar A, Hohenstein EG, Pritchard BP, Brooks BR, Schaefer HF 3rd, Sokolov AY, Patkowski K, DePrince AE 3rd, Bozkaya U, King RA, Evangelista FA, Turney JM, Crawford TD, Sherrill CD, Psi4 1.4: Open-source software for high-throughput quantum chemistry, J Chem Phys. 2020 May 14;152(18):184108. doi: 10.1063/5.0006002. PMID: 32414239; PMCID: PMC7228781, https://www.ncbi.nlm.nih.gov/pmc/articles/PMC7228781/pdf/JCPSA6-000152-184108_1.pdf

[PvsNP] P and NP, www.cs.uky.edu. Archived from the original on 2016-09-19, https://web.archive.org/web/20160919023326/http://www.cs.uky.edu/~lewis/cs-heuristic/text/class/p-np.html

[PyQMC] PyQMC, a Python module that implements real-space quantum Monte Carlo techniques, https://github.com/WagnerGroup/pyqmc

[PySCF] The Python-based Simulations of Chemistry Framework (PySCF), https://pyscf.org/

[Q Daily] Quantum Technology | Our Sustainable Future, The Quantum Daily, Jul 29, 2021, https://www.youtube.com/watch?v=iB2_ibvEcsE

[QC40] (Livestream) QC40: Physics of Computation Conference 40th Anniversary, https://www.youtube.com/watch?v=GR6ANm6Z0yk

[QMC] Google Quantum AI, Unbiased fermionic Quantum Monte Carlo with a Quantum computer, Quantum Summer Symposium 2021, 30 July 2021, https://www.youtube.com/watch?v=pTHtyKuByvw

[Qa_Zoo] Stephen Jordan, Algebraic and Number Theoretic Algorithms, https://quantumalgorithmzoo.org/

[Qiskit] Qiskit, https://qiskit.org/

[Qiskit_2021_Lab4] Julien Gacon, Lab 4: Introduction to Training Quantum Circuits, Qiskit Summer School 2021, https://learn.qiskit.org/summer-school/2021/lab4-introduction-training-quantum-circuits

[Qiskit_Alg] Linear Algebra, Qiskit, https://qiskit.org/textbook/ch-appendix/linear_algebra.html

[Qiskit_Nat_0] Qiskit_Nature, https://github.com/Qiskit/qiskit-nature/blob/main/README.md

[Qiskit_Nat_1] Qiskit Nature & Finance Demo Session with Max Rossmannek & Julien Gacon, Oct 15, 2021, `https://www.youtube.com/watch?v=UtMVoGXlz04`

[Qiskit_Nat_2] FermionicOp, `https://qiskit.org/documentation/nature/stubs/qiskit_nature.operators.second_quantization.FermionicOp.html`

[Qiskit_Nat_3] ElectronicStructureProblem.second_q_ops, `https://qiskit.org/documentation/nature/stubs/qiskit_nature.problems.second_quantization.ElectronicStructureProblem.second_q_ops.html`

[Qiskit_Nat_4] QubitConverter, `https://qiskit.org/documentation/nature/stubs/qiskit_nature.converters.second_quantization.QubitConverter.html`

[Qiskit_Nat_5] Qiskit Nature Tutorials, Electronic structure, `https://qiskit.org/documentation/nature/tutorials/01_electronic_structure.html`

[Qiskit_Nat_6] Qiskit Nature Tutorials, Sampling the potential energy surface, `https://qiskit.org/documentation/nature/_modules/qiskit_nature/algorithms/pes_samplers/bopes_sampler.html`

[Qiskit_Nat_T] Second-Quantization Operators (qiskit_nature.operators.second_quantization) > FermionicOp > FermionicOp.set_truncation, `https://qiskit.org/documentation/nature/stubs/qiskit_nature.operators.second_quantization.FermionicOp.set_truncation.html`

[Qiskit_Nature] Introducing Qiskit Nature, Qiskit, Medium, April 6, 2021, `https://medium.com/qiskit/introducing-qiskit-nature-cb9e588bb004`

[QuTiP] QuTiP, Plotting on the Bloch Sphere, `https://qutip.org/docs/latest/guide/guide-bloch.html`

[Quantum_AI] The Quantum AI Sustainability Symposium, Q4Climate, Speakers: Dr. Karl Thibault, Mr. Michał Stęchły, Sep 1, 2021, `https://quantum.ieee.org/conferences/quantum-ai`

[Rayleigh_Ritz] Rayleigh-Ritz method, Wikipedia, `https://en.wikipedia.org/wiki/Rayleigh%E2%80%93Ritz_method`

[Ribeiro] Sofia Leitão, Diogo Cruz, João Seixas, Yasser Omar, José Emilio Ribeiro, J.E.F.T. Ribeiro, Quantum Simulation of Fermionic Systems, CERN, `https://indico.cern.ch/event/772852/contributions/3505906/attachments/1905096/3146117/Quantum_Simulation_of_Fermion_Systems.pdf`

[Rioux] Mach-Zehnder Polarizing Interferometer Analyzed Using Tensor Algebra, https://faculty.csbsju.edu/frioux/photon/MZ-Polarization.pdf

[Rydberg_R] Rydberg constant, from Wikipedia, https://en.wikipedia.org/wiki/Rydberg_constant

[Rydberg_Ritz] Rydberg-Ritz combination principle, Wikipedia, https://en.wikipedia.org/wiki/Rydberg%E2%80%93Ritz_combination_principle

[Sawaya] Nicolas P. D. Sawaya, Francesco Paesani, Daniel P. Tabor, Near- and long-term quantum algorithmic approaches for vibrational spectroscopy, 1 Feb 2021, arXiv:2009.05066 [quant-ph], https://arxiv.org/abs/2009.05066

[Schiffer_1] Fabijan Pavošević, Tanner Culpitt, and Sharon Hammes-Schiffer, Multicomponent Quantum Chemistry: Integrating Electronic and Nuclear Quantum Effects via the Nuclear–Electronic Orbital Method, Chem. Rev. 2020, 120, 9, 4222–4253, https://doi.org/10.1021/acs.chemrev.9b00798

[Schiffer_2] Kurt R. Brorsen, Andrew Sirjoosingh, Michael V. Pak, and Sharon Hammes-Schiffer, Nuclear-electronic orbital reduced explicitly correlated Hartree-Fock approach: Restricted basis sets and open-shell systems, J. Chem. Phys. 142, 214108 (2015), https://doi.org/10.1063/1.4921304

[SciPy] Statistical functions (scipy.stats), https://docs.scipy.org/doc/scipy/getting_started.html

[SciPy_0], SciPy, https://scipy.org/

[SciPy_sph] SciPy, API reference, Compute spherical harmonics, scipy.special.sph_harm, https://docs.scipy.org/doc/scipy/reference/generated/scipy.special.sph_harm.html

[Seeley] Jacob T. Seeley, Martin J. Richard, Peter J. Love, The Bravyi-Kitaev transformation for quantum computation of electronic structure, 29 Aug 2012, arXiv:1208.5986 [quant-ph], https://arxiv.org/abs/1208.5986v1

[Sharkey] K. Sharkey et. al., Non-Born-Oppenheimer method for direct variational calculations of diatomic first excited rotational states using explicitly correlated all-particle Gaussian functions, Table I. Total energies (in hartrees) of the (v,0) and (v,1) states of H_2., Physical Review A, 88, 032513 (2013), https://journals.aps.org/pra/abstract/10.1103/PhysRevA.88.032513

[Sharkey_0] Keeper L. Sharkey and Ludwik Adamowicz, An algorithm for nonrelativistic quantum mechanical finite-nuclear-mass variational calculations of nitrogen atom in L = 0, M = 0 states using all-electrons explicitly correlated Gaussian basis functions, J. Chem. Phys. 140, 174112 (2014), https://doi.org/10.1063/1.4873916

[Sharkey_1] K. Sharkey, Molecular-hydrogen poster, QLEAN™, https://qlean.world/molecular-hydrogen-poster

[Shor] Peter Shor, The Story of Shor's Algorithm, Straight From the Source, July 2, 2021, https://www.youtube.com/watch?v=6qD9XElTpCE

[Skylaris] C.-K. Skylaris, CHEM6085: Density Functional Theory, Lecture 8, Gaussian basis sets, https://www.southampton.ac.uk/assets/centresresearch/documents/compchem/DFT_L8.pdf

[Skylaris_1] C.-K. Skylaris, CHEM3023: Spins, Atoms and Molecules, Lecture 8, Experimental observables / Unpaired electrons, https://www.southampton.ac.uk/assets/centresresearch/documents/compchem/chem3023_L8.pdf

[Sph_Real] Wikipedia, Spherical Harmonics, Real forms, https://en.wikipedia.org/wiki/Spherical_harmonics#Real_forms

[Spheres] How to Prepare a Permutation Symmetric Multiqubit State on an Actual Quantum Computer, https://spheres.readthedocs.io/en/stable/notebooks/9_Symmetrized_Qubits.html

[Stephens] Matthew Stephens, The Metropolis Hastings Algorithm, https://stephens999.github.io/fiveMinuteStats/MH_intro.html

[Stickler] B. A. Stickler et al., Enantiomer superpositions from matter-wave interference of chiral molecules, Phys. Rev. X 11, 031056 (2021), https://journals.aps.org/prx/abstract/10.1103/PhysRevX.11.031056

[Su] Yuan Su, Dominic W. Berry, Nathan Wiebe, Nicholas Rubin, and Ryan Babbush, Fault-Tolerant Quantum Simulations of Chemistry in First Quantization, 11 Oct. 2021, PRX Quantum 2, 040332, DOI:10.1103/PRXQuantum.2.040332, https://doi.org/10.1103/PRXQuantum.2.040332

[SymPy] SymPy, A Python library for symbolic mathematics, https://www.sympy.org/en/index.html

[SymPy_CG] SymPy, Clebsch-Gordan Coefficients, https://docs.sympy.org/latest/modules/physics/quantum/cg.html

[SymPy_Rnl] Hydrogen Wavefunctions, https://docs.sympy.org/latest/modules/physics/hydrogen.html

[Tamir] Abraham Tamir, Applications of Markov Chains in Chemical Engineering, Elsevier, 1998, 9780080527390, 0080527396, https://www.google.fr/books/edition/Applications_of_Markov_Chains_in_Chemica/X0ivOmHYPoYC

[Thibault] Casey Berger, Agustin Di Paolo, Tracey Forrest, Stuart Hadfield, Nicolas Sawaya, Michał Stęchły, Karl Thibault, Quantum technologies for climate change: Preliminary assessment, IV. ENERGY EFFICIENCY OF QUANTUM COMPUTERS By Karl Thibault, arXiv:2107.05362 [quant-ph], 23 Jun 2021, https://arxiv.org/abs/2107.05362

[Toulouse] Julien Toulouse, Introduction to quantum chemistry, Jan 20, 2021, https://www.lct.jussieu.fr/pagesperso/toulouse/enseignement/introduction_qc.pdf

[Toulouse_1] Julien Toulouse, Quantum Monte Carlo wave functions and their optimization for quantum chemistry, CEA Saclay, SPhN Orme des Merisiers, April 2015, https://www.lct.jussieu.fr/pagesperso/toulouse/presentations/presentation_saclay_15.pdf

[Tranter] Andrew Tranter, Peter J. Love, Florian Mintert, Peter V. Coveney, A comparison of the Bravyi-Kitaev and Jordan-Wigner transformations for the quantum simulation of quantum chemistry, 5 Dec 2018, J. Chem. Theory Comput. 2018, 14, 11, 5617–5630, https://doi.org/10.1021/acs.jctc.8b00450

[Troyer] Matthias Troyer: Achieving Practical Quantum Advantage in Chemistry Simulations, QuCQC 2021, https://www.youtube.com/watch?v=2MsfbPlKgyI

[Ucsd] University of Californian San Diego, Spherical Coordinates and the Angular Momentum Operators, https://quantummechanics.ucsd.edu/ph130a/130_notes/node216.html

[VQE_1] Peruzzo, A., McClean, J., Shadbolt, P. et al., A variational eigenvalue solver on a photonic quantum processor, Nat Commun 5, 4213 (2014), https://doi.org/10.1038/ncomms5213

[VQE_2] Qiskit Nature, Ground state solvers, https://qiskit.org/documentation/nature/tutorials/03_ground_state_solvers.html

[VQE_3] Hardware-efficient variational quantum eigensolver for small molecules and quantum magnets, Nature 549, 242–246 (2017), https://doi.org/10.1038/nature23879

[VQE_4] Running VQE on a Statevector Simulator, `https://qiskit.org/textbook/ch-applications/vqe-molecules.html#Running-VQE-on-a-Statevector-Simulator`

[Vandersypen] Vandersypen, L., Steffen, M., Breyta, G. et al., Experimental realization of Shor's quantum factoring algorithm using nuclear magnetic resonance, Nature 414, 883–887 (2001), `https://doi.org/10.1038/414883a`

[Veis] Libor Veis, Jakub Višňák, Hiroaki Nishizawa, Hiromi Nakai, Jiří Pittner, Quantum chemistry beyond Born–Oppenheimer approximation on a quantum computer: A simulated phase estimation study, International Journal of Quantum Chemistry, 22 June 2016, `https://doi.org/10.1002/qua.25176`

[Wiki-Comb] Number of k-combinations for all k, Wikipedia, `https://en.wikipedia.org/wiki/Combination#Number_of_k-combinations_for_all_k`

[Wiki-GAU] Gaussian orbital, Wikipedia, `https://en.wikipedia.org/wiki/Gaussian_orbital`

[Wiki-STO] Slater-type orbital, Wikipedia, `https://en.wikipedia.org/wiki/Slater-type_orbital`

[Wiki_1] Mathematical formulation of quantum mechanics, Wikipedia, `https://en.wikipedia.org/wiki/Mathematical_formulation_of_quantum_mechanics`

[Wonders] Optical Isomers, Enantiomers and Chiral Molecules, WondersofChemistry, `https://www.youtube.com/watch?v=8TIZdWR4gIU`

[Yepez] Jeffrey Yepez, Lecture notes: Quantum gates in matrix and ladder operator forms, Jan 15, 2013, `https://www.phys.hawaii.edu/~yepez/Spring2013/lectures/Lecture2_Quantum_Gates_Notes.pdf`

9
Glossary

By convention an asterisk indicates that there is a complimentary entry in *Appendix A – Readying Mathematical Concepts*.

Angular momentum quantum number *

Also known as the orbital quantum number or the azimuthal quantum number, and denoted by l, this describes the electron subshell and gives the magnitude of the orbital angular momentum.

Anti-commutator *

An operation of two operators A, B defined as: $\{A, B\} \overset{\text{def}}{=} AB + BA$.

Anti-commutation *

A set of fermionic annihilation operators and creation operators that act on local electron modes can be defined that satisfy anti-commutation relations.

Atom

A basic particle that composes a chemical element. It consists of a nucleus surrounded by moving electrons arranged in orbitals that describe their positions in terms of probabilities. The atom has no overall electric charge.

Atomic number

A number that identifies an atom by the number of protons/electrons in the atom being specified.

Atomic or molecular orbital

A mathematical function that describes the probability of finding an electron at a given point in space at a given time in an atom or molecule.

Atomic Simulation Environment (ASE)

A set of tools and Python modules for setting up, manipulating, running, visualizing, and analyzing atomistic simulations.

Basis set

A set of functions that are combined in linear combinations to create molecular orbits.

Born-Oppenheimer (BO) approximation

This is the assumption that the wave functions of atomic nuclei and electrons in a molecule can be treated separately such that the electronic motion and the nuclear motion can be separated.

Bravyi-Kitaev (BK) transformation

A transformation method of mapping the occupation state of a fermionic system onto qubits. This transformation maps the Hamiltonian of n interacting fermions to an $O(\log n)$ local Hamiltonian of n qubits.

Center-of-mass (COM)

Mean location of a distribution of mass, which is the average location weighted by the masses of the elements in a many-body system.

Clebsch-Gordon (CG) coefficients

The expansion coefficients of coupled total angular and/or spin momentum in an uncoupled tensor product basis.

Commutator *

An operation of two operators A, B defined as: $[A, B] \stackrel{\text{def}}{=} AB - BA$. For any operators A and B, $[A, B] = 0$ if and only if A and B commute. It can be shown that if a quantum system has two simultaneously physically observable quantities, then the Hermitian operators which represent them must commute.

Complete

For a function where all statistically important data that are needed to represent that quantum system is available such that calculations of properties converge to a limit, that is, a single value.

Constructive interference

When two or more waves add together so that the amplitude of the resulting wave is equal to the sum of the individual amplitudes.

Coupled-cluster (CC)

A theory that constructs a multi-electron wavefunction using the exponential cluster operator, which is the sum of the operator for all single excitations, the operator for all double excitations, and so on.

Density-functional theory (DFT)

A simulation method based on quantum mechanical first principles (ab initio) and spatially depend on density functionals that describe the electronic structure properties of atomic systems, atoms, molecules, and crystals.

Density matrix *

Describes the state of a quantum system based on probabilities, average value, and outcome for the measurement performed.

Destructive interference

When the maxima of two waves are out of phase by π radians, the resulting wave has smaller amplitude and can be zero amplitude, and this effect is called destructive interference.

Dirac notation *

Dirac notation is also known as bra-ket notation. The state of a quantum system, or the wave function, is represented by a ket, which is a column vector of coordinates and/or variables. The bra denotes a linear function that maps each column vector to a complex conjugate row vector. The action of the row vector on a column vector.

Eigenvalue

When the result of applying a linear transformation to a vector is that vector multiplied by a scalar, then the vector is called an eigenvector, and the scalar is called the eigenvalue associated with that linear transformation.

Electron

A stable subatomic particle that has a negative electric charge, is a component of all atoms, and is the primary carrier of electricity in solids.

Electronic structure molecular Hamiltonian

The Hamiltonian operator for a molecule represents the total energy of all its particles, electrons, and nuclei, comprising the sum of the kinetic energy of all particles and the potential energy between all particles.

Entangled

If the wave function for a multiple-particle system cannot be factored into a product of single-particle functions, then the quantum system is said to be entangled.

Exchange operator

This is an operator that permutes the unphysical labels of the particles. It acts on states in Fock space and identifies if identical particles are bosons or fermions.

Expectation value

This is the sum of all the possible outcomes of a measurement of a state weighted by their probabilities.

Explicitly correlated Gaussian (ECG)

This is the square of the distance between all pairs of particles in an exponential form.

Explicitly correlated Gaussian basis functions (ECGs)

More than one ECG in a set.

Fermion, fermionic, electron annihilation operator *

A mathematical operation that allows us to represent excitations or transitions of quasi particles. An excitation requires the initial state to be at a lower energy level than the final state.

Fermion, fermionic, electron creation operator *

A mathematical operation that allows us to represent disexcitement (de-exitation) or transitions of quasi particles. A de-excitation requires the initial state to be at a higher energy level than the final state.

Fermion, fermionic, electron excitation operator *

A mathematical operation that excites an electron from an occupied spin orbital into an unoccupied orbital.

Fock space

An algebraic construct used in quantum mechanics to create the quantum states space of a variable or unknown number of identical particles from the Hilbert space of a single particle.

Gaussian-type orbitals (GTOs)

Functions that are utilized as atomic orbitals in the Linear Combination of Atomic Orbitals (LCAO) method to represent electron orbitals in molecules and a variety of attributes that are dependent on them.

Gradient descent

An optimization algorithm for finding a local minimum of a function by iteratively moving in the opposite direction of the gradient of the function at a given point that is the direction of steepest descent. This can be compared to a ball rolling around the landscape. If we give the ball a nudge in some direction, generally downward, the ball will end up in the minimum.

Hamiltonian operator

The operator associated with the total energy of the quantum system, and it is a sum of potential and kinetic energy operators.

Hartree (Ha)

An atomic unit of energy commonly used in molecular orbital calculations which is defined as $2R_\infty hc$, where R_∞ is the Rydberg constant, h is the Planck constant, and c is the speed of light.

Hartree-Fock method

A technique that is used as an approximation approach for determining the wave function and energy of a quantum many-body system in a stationary state.

Hermitian operator

A linear operator that is equal to its transpose conjugate, that is, self-adjoint, and has real eigenvalues that correspond to an observable.

Jordan-Wigner transformation

The Jordan-Wigner transformation is widely used to simulate a system of electrons with the same number of qubits as electrons. It stores the occupation of each spin orbital in each qubit and maps the fermionic creation and annihilation operators to the tensor product of Pauli operators.

Linear Combination of Atomic Orbitals (LCAO)

A superposition of atomic orbitals.

Magnetic quantum number

Describes the electron's energy level within its subshell and the orientation of the electron's orbital and is denoted by m_l. It can take on integer values ranging from $-l, ..., 0, ..., +l$ where l is the angular momentum quantum number.

Markov chain Monte Carlo (MCMC)

A method for producing samples from a target probability distribution by simulating a Markov chain whose stationary distribution is the target probability distribution.

Markov Chain theory (MCT)

An approximation method of a (non-)quantum system that can occupy various states, and whose (time-)evolution is defined once an initial state and the probability transitions between the states are fixed.

Mean-field theory (MFT)

An approximation method that reduces the many-body interactions of a system by one effective interaction with a mean-field.

Metropolis-Hastings (MH)

A Markov chain Monte Carlo (MCMC) method for producing samples from a probability.

Mixed quantum state

A statistical ensemble of pure quantum states.

Molecular Hamiltonian

The total energy operator for a molecule representing its particles, comprising the sum of the kinetic energy of all particles and the potential energy between all particles.

Molecular Orbital (MO) theory

A method for approximating the wave functions of electrons in a molecule, the molecular orbitals, as linear combinations of atomic orbitals (LCAO) by applying the density functional theory (DFT) or the Hartree-Fock method.

Monte Carlo method

Any stochastic method used to solve a problem.

Neutron

A nucleon that has an electric charge of zero and a mass that is 1,838.68366173 times greater than the electron. The neutron is slightly heavier than the proton.

Normalizable

A wave function that is a solution of the Schrödinger equation such that the integral of its squared module is finite, meaning that the positive definite product should be less than infinity when integrated over all space.

Nucleons

The building blocks of atomic nuclei, which are protons and neutrons.

Nucleus (plural nuclei)

The core of every atom contains one or more protons and zero or more neutrons, which are held together by the strong nuclear force.

Numerical Python (NumPy)

An open-source Python library that is used in almost every field of science and engineering.

Occupation number

In a basis of the Fock space, the occupation number pertaining to a spin-orbital state is 0 if the spin-orbital state is not occupied by an electron and 1 if it is occupied by an electron.

Occupation number operator *

An operator that acts on local electron modes and satisfy anti-commutation relations.

Occupation number representation

A synonym of the second quantization representation.

Occupied spin-orbital

A spin-orbital that is occupied by an electron.

Parity transformation

Dual to the Jordan-Wigner transformation, the parity operators are low-weight, while the occupation operators become high-weight.

Pauli exclusion principle (PEP) *

In 1925, Pauli described the PEP for electrons, which states that it is impossible for two electrons of the same atom to simultaneously have the same values of the following four quantum numbers: the principal quantum number, the angular momentum quantum number, the magnetic quantum number, and the spin quantum number.

Following the discovery of various types of elementary particles, the PEP for electrons has been generalized for all elementary particles and composite systems. For fermions, the total wave function must be antisymmetric with respect to the exchange of identical pair particles. For bosons, the total wave function must be symmetric with respect to the exchange of pair particles. For composite systems with both identical fermions and identical bosons, the preceding operations must hold true simultaneously.

Pauli Matrices

A set of three 2×2 complex matrices that are Hermitian, unitary, and represent the orbital and spin angular momentum magnetic interactions.

Polymerase chain reaction (PCR)

An amplification technique to replicate complex proteins using a series of the same chemical reaction.

Potential energy surface (PES)

The potential energy of a system, usually a molecule, that describes a function of parameters such as the bond length and the bond angle between two atoms.

Positive definite

This applies to matrices, vectors, and wave functions such that the complex conjugate transpose, indicated by a dagger (†), multiplied by itself, is strictly greater than zero.

Principal quantum number

Describes the energy level or the electron's position in a shell of the atom and is numbered from one up to the shell containing the outermost electron of that atom that can range from one to infinity, thus it is a continuous quantum number. However, as the electron is excited to higher and higher values, and dissociates from the atom, it is then considered a free electron plus an ion. This process is called ionization, and the energy levels are considered to be discrete.

Proton

A stable subatomic particle that is a component of a nucleus of an atom and carries a positive electric charge equal in magnitude to that of an electron and a mass that is 3 orders of magnitude higher than the electron.

Pseudopotential

An effective potential that replaces the full atomic all-electron potential, eliminates the core states, and describes the valence electrons by pseudo-wavefunctions.

Pure quantum state

A synonym for a state vector or a wave function.

Python-based Simulations of Chemistry Framework (PySCF)

An open-source collection of electronic structure modules powered by Python.

Quantum chemistry

A branch of chemistry that aims to understand chemical systems starting from the postulates of quantum mechanics.

Quantum mechanics

A fundamental physics theory that mathematically describes the behavior of matter, energy, and the interactions with light at the scale of subatomic particles, atoms, and molecules.

Quantum Monte Carlo (QMC)

A Monte Carlo method applied to a quantum system.

Quantum Phase Estimation (QPE) *

An algorithm to measure the phase of a quantum state.

Quantum Toolbox in Python (QuTiP)

A general framework for solving quantum mechanics problems such as systems composed of few-level quantum systems and harmonic oscillators.

Qiskit

An open-source software development kit (SDK) for working with quantum computers at the level of pulses, circuits, and application modules.

Quantum Natural SPSA (QN-SPSA)

A gradient descent method used to optimize systems that is based on SPSA and samples the natural gradient instead of the first gradient by approximating Hessian of the fidelity of the ansatz circuit.

Qubit

A quantum bit of computing information that is represented by a state vector through coupling together angular and spin momentum.

Qubit Hamiltonian

A Hermitian operator that is represented by a Hermitian matrix.

Rayleigh-Ritz variational theorem

The expectation value of the Hamiltonian of a system is always an upper bound to the lowest energy associated with the eigenvalue being solved for.

Restricted HF (RHF) method

A HF method used for closed-shell molecules. The spin-orbitals are either alpha (spin-up) or beta (spin-down) and all orbitals are doubly occupied by alpha and beta spin-orbitals.

Restricted open-shell (ROHF) method

A HF method used for open-shell molecules where the numbers of electrons of each spin are not equal. ROHF uses as many doubly occupied molecular orbitals as possible and singly occupied orbitals for the unpaired electrons.

Second quantization representation

A representation of quantum many-body states in the Fock-state basis.

Sequential Least Squares Programming (SLSQP)

A sequential quadratic programming optimization algorithm originally defined by Dieter Kraft.

Simultaneous Perturbation Stochastic Approximation (SPSA)

A gradient descent method used to optimize systems that uses the stochastic gradient approximation and only performs two measurements of the objective function at each step.

SymPy

A Python library for symbolic mathematics.

Single-valued

A function that, for a given input variable, only has one possible output.

Slater determinant wave function

An expression that describes the wave function of a multi-fermionic system satisfying anti-symmetry criteria of the PEP by changing sign when two electrons are exchanged.

Slater-type orbitals (STOs)

Functions used in a determinate to formulate atomic orbitals and molecular orbitals through a linear combination of atomic orbitals (LCAO).

Spin projection quantum number

Gives the projection of the spin momentum s along the specified axis as either spin up (+½) or spin down (-½) in a given spatial direction, which in quantum computing is defined as the z-axis.

Spin quantum number

Describes the intrinsic spin momentum of a certain particle type; it varies for each particle type, and there is no classical analog to describe what it is. For the electron, it is equal to 1/2.

Square integrable

A real or complex-valued function for which the square of the absolute value is finite for the integration over all possible values of the domain.

State vector

A vector used to represent the wave function of a quantum state.

Superposition

A linear combination of all real or complex basis functions.

Total wave function *

Describes the physical behavior of a system. It contains all the information of a quantum system including complex numbers as parameters. In general, it is a function of all the particles in the system and includes the spatial position, the spin directional coordinates of each particle, and time.

Trotterization

A truncation method of quantum simulations that is widely used to simulate non-commuting Hamiltonians on quantum computers.

Unitary Coupled Cluster Singles and Doubles (UCCSD)

A unitary coupled-cluster theory that constructs a multi-electron wavefunction using the exponential cluster operator, which is the sum of the operator for all single excitations and of the operator for all double excitations.

Unrestricted HF (UHF) method

A HF method used for open-shell molecules where the numbers of electrons of each spin are not equal. UHF orbitals can have either alpha or beta spin, but the alpha and beta orbitals may have different spatial components.

Variational Monte Carlo (VMC)

A Quantum Monte Carlo method that implements a variational method.

Variational Quantum Algorithm (VQA)

An algorithm which uses a parameterized quantum circuit to prepare a trial quantum state as a trial solution (an ansatz) and a classical computer to optimize the parameters of this quantum circuit with respect to an objective function.

Variational Quantum Eigensolver (VQE)

An algorithm introduced in 2014 that is defined using quantum-based hardware. It is the first of several variational quantum algorithms (VQAs) that are currently being explored by the scientific industry. In a loop, a classical computer optimizes the parameters of a quantum circuit with respect to an objective function, such as finding the ground state of a molecule, which is the state with the lowest energy. The parameterized quantum circuit prepares a trial quantum state as a trial solution (an ansatz). By repeatedly measuring qubits at the output of the quantum circuit, we get the expectation value of the energy observable with respect to the trial state.

Appendix A
Readying Mathematical Concepts

By convention, a # indicates that there is a complimentary entry in *Chapter 9, Glossary*.

In this appendix, we will cover the following topics:

- Notations used
- Mathematical definitions

Technical requirements

A companion Jupyter notebook for this chapter can be downloaded from GitHub at `https://github.com/PacktPublishing/Quantum-Chemistry-and-Computing-for-the-Curious`, which has been tested in the Google Colab environment, which is free and runs entirely in the cloud, and in the IBM Quantum Lab environment. Please refer to *Appendix B – Leveraging Jupyter Notebooks in the Cloud*, for more information. The companion Jupyter notebook automatically installs the following list of libraries:

- **Numerical Python** (**NumPy**) [NumPy], an open-source Python library that is used in almost every field of science and engineering
- **SymPy** [SymPy], a Python library for symbolic mathematics
- **Qiskit** [Qiskit], an open-source SDK for working with quantum computers at the level of pulses, circuits, and application modules
- Qiskit visualization support to enable the use of its visualization functionality and Jupyter notebooks

Installing NumPy, SimPy, and Qiskit and importing various modules

Install NumPy with the following command:

```
pip install numpy
```

Install SymPy using the following command:

```
pip install simpy
```

Install Qiskit using the following command:

```
pip install qiskit
```

Importing NumPy and a function that returns a LaTeX representation of a complex array

Import NumPy with the following command:

```
import numpy as np
```

Import the required functions and class methods. The `array_to_latex function()` returns a LaTeX representation of a complex array with dimension 1 or 2:

```
from qiskit.visualization import array_to_latex
```

Notations used

We will be using the following notations wherever it is appropriate:

- α, β, θ, σ, φ, and so on – Lowercase Greek letters for scalars.

- a, b, x, and so on – Lowercase Latin letters for column vectors in particle space. These vectors have n components denoted a_k, b_k, x_k, and so on where k is an integer.

- A, B, X, and so on – Uppercase Latin letters for matrices in particle space. These are n X n matrices.

- a', a^T, A', A^T, and so on – The prime symbol (') and the letter T stand for vector and matrix transpose.

- a^*, X^*, and so on – The asterisk symbol (*) is used for vector and matrix complex conjugate.

- $x^\dagger$, $X^\dagger$, and so on – The dagger symbol, $\dagger$, is used for vector and matrix complex conjugate transpose.

- A^{-1}, and so on – A power to the negative one $(^{-1})$ represents the inverse of a matrix.

- $X \otimes Y$, and so on – The o-times symbol $\otimes$ represents the Kronecker product or tensor product of matrices and/or vectors.

- $X \oplus Y$, and so on. – The symbol $\oplus$ represents the Kronecker sum of square matrices.

- $\exists$ – There exists at least one.

- $\forall$ – For all.

- $\in$ – Is member of, for example $\alpha \in \mathbb{R}$ means that α is in the set $\mathbb{R}$ of real numbers.

Mathematical definitions

Pauli exclusion principle (PEP)

In 1925, Pauli described the PEP for electrons, which states that it is impossible for two electrons of the same atom to simultaneously have the same values of the following four quantum numbers: n, the principal quantum number; l, the angular momentum quantum number; m_l, the magnetic quantum number; and m_s, the spin quantum number.

Following the discovery of various types of elementary particles, the PEP for electrons has been generalized for all elementary particles and composite systems. Remember that fermions are particles that have half-integer spin ($s = \frac{1}{2}, \frac{3}{2}, \frac{5}{2}, ...$) and bosons are particles that have integer spin ($s = 0, 1, 2, ...$). The general formulation of the PEP states the total wave function Ψ_{total} for a quantum system must have certain symmetries for all sets of identical particles, that is, electrons and identical nuclei, both bosons and fermions, under the operation of pair particle permutation:

- For fermions, the total wave function must be antisymmetric ($-$) with respect to the exchange of identical pair particles ($\hat{A}_{ij}$):

$$\hat{A}_{ij}\Psi_{total} = -\Psi_{total}$$

meaning that the spatial part of the wave function is antisymmetric while the spin part is symmetric, or vice versa.

- For bosons, the total wave function must be symmetric ($+$) with respect to the exchange of pair particles ($\hat{S}_{ij}$):

$$\hat{S}_{ij}\Psi_{total} = +\Psi_{total}$$

meaning that both the spatial wave function and spin function are symmetric, or both are antisymmetric.

- For composite systems with both identical fermions and identical bosons, the preceding operations must hold true simultaneously.

Angular momentum quantum number

Also known as the orbital quantum number or the azimuthal quantum number, and denoted by l, this describes the electron subshell and gives the magnitude of the orbital angular momentum through the relation: $L^2 = \hbar^2 l(l + 1)$. In chemistry and spectroscopy, $l = 0$ is called the s orbital, $l = 1$ the p orbital, $l = 2$ the d orbital, and $l = 3$ the f orbital. Technically, there are more orbitals beyond the f orbital, that is, $l = 4 = g, l = 5 = h$, and so on, and these are of higher energy levels.

Occupation number operator

An operator $\hat{a}_j^\dagger \hat{a}_j$ where $j \in [0, N - 1]$, and $\{\hat{a}_i\}_{i=0}^{N-1}$ are the annihilation operators and $\{\hat{a}_j^\dagger\}_{j=0}^{N-1}$ are the creation operators that act on local electron modes, and satisfy the following anti-commutation relations:

$$\{\hat{a}_i, \hat{a}_j^\dagger\} = \hat{a}_i^\dagger \hat{a}_j + \hat{a}_j \hat{a}_i^\dagger = \delta_{ij} = \begin{cases} 0, & i \neq j \\ 1, & i = j \end{cases}$$

$$\{a_i^\dagger, a_j^\dagger\} = \{\hat{a}_i, \hat{a}_j\} = 0$$

where δ_{ji} is the Dirac delta function and $\{A, B\} \stackrel{\text{def}}{=} AB + BA$ is the anti-commutator of two operators A and B.

Quantum Phase Estimation (QPE)

Given a unitary operator U, its eigenstate and eigenvalues, $U|\psi\rangle = e^{2\pi i \theta}|\psi\rangle$, the ability to prepare a state $|\psi\rangle$, and the ability to apply U itself, the QPE algorithm calculates $2^n \theta$, where n is the number of qubits used to estimate θ thereby allowing measurement of θ as precisely as we want.

Complex numbers

Complex numbers are of the form $\varphi = \alpha + i\beta$ where α and β are real numbers and i (j in Python) is called the imaginary unit that by definition satisfies the equation $i^2 = -1$. The magnitude of a complex number is: $|\varphi| = |\alpha + i\beta| = \sqrt{\alpha^2 + \beta^2}$. The complex conjugate of $\varphi = \alpha + i\beta$ is $\varphi^* = \alpha - i\beta$. Euler's formula $e^{i\theta} = \cos\theta + i\sin\theta$ is convenient for multiplying complex numbers, and exponentiation. The set of complex numbers together with the addition and multiplication operations is a field denoted $\mathbb{C}$. Algebraic expressions composed of complex numbers follow the standard rules of algebra; the difference with real numbers is that i^2 is replaced by -1.

Vector space

A vector space V over the field $\mathbb{C}$ of complex numbers or the field R of real numbers is a set of objects called vectors that can be added together and multiplied ("scaled") by numbers.

The following Python code illustrates a vector with two complex components:

```
x = np.array([[1j],
              [2]])
array_to_latex(x, prefix='x = ')
```

$$x = \begin{bmatrix} i \\ 2 \end{bmatrix}$$

We use the @ operator introduced in Python 3.5 to multiply a vector by a number, as illustrated below, multiplying vector $x = \begin{bmatrix} i \\ 2 \end{bmatrix}$ by the imaginary unit i (j in Python) where i^2 is replaced by -1:

```
α = 1j
print('α =', α)
y = α*x
array_to_latex(y, prefix=' y = α*x =')
```

$$\alpha = 1j$$
$$y = \alpha * x = \begin{bmatrix} -1 \\ 2i \end{bmatrix}$$

Linear operators

A function f defined on a vector space V over $\mathbb{C}$ is a linear operator if it has the two following properties:

- For any x, y in V, $f(x + y) = f(x) + f(y)$
- For any x in V, α in $\mathbb{C}$, $f(\alpha x) = \alpha f(x)$

Matrices

A matrix is a set of elements arranged in a square or rectangular array. Elements can be numbers, matrices, functions, or algebraic expressions. The order or shape of a matrix is written (number of rows) x (number of columns). Indices are written in *row, column* format so for example, $a_{k,l}$ is an element in row k and column l. Matrices represent linear operators in a vector space. It is convenient to use the same symbol for an operator and its matrix in some orthonormal basis.

Eigenvalues and eigenvectors

By definition, an eigenvector of a linear operator f defined on a vector space V over $\mathbb{C}$ is a non-zero vector x that has the following property: $f(x) = \lambda x$ where λ is a scalar in $\mathbb{C}$ known as the eigenvalue associated with the eigenvector x.

For a finite-dimensional space V, the above definition is equivalent to $Ax = \lambda x$ where A is the matrix representation of f.

Vector and matrix transpose, conjugate, and conjugate transpose

The transpose of some vector a or some matrix A often denoted as a', a^T, A', A^T is obtained by switching the row and column indices of the vector a or matrix A. The following Python code illustrates the transpose of vector $x = \begin{bmatrix} i \\ 2 \end{bmatrix}$:

```
x = array_to_latex(x.transpose(), prefix='x^T = ')
```

$$x^T = \begin{bmatrix} i & 2 \end{bmatrix}$$

The complex conjugate of some vector a or some matrix A often denoted as a^*, A^* is obtained by performing the complex conjugate of all the elements:

```
x = array_to_latex(x.conjugate(), prefix='x^* = ')
```

$$x^* = \begin{bmatrix} -i \\ 2 \end{bmatrix}$$

The complex conjugate transpose of some vector a or matrix A often is denoted as $a^\dagger$, $A^\dagger$ in quantum mechanics. The symbol, $\dagger$, is called the dagger. $A^\dagger$ is called the adjoint or Hermitian conjugate of A:

```
x = array_to_latex(x.conjugate().transpose(), prefix='(x^*)^T =
')
```

$$x^\dagger = (x^*)^T = \begin{bmatrix} -i & 2 \end{bmatrix}$$

Dirac's notation

In Dirac's notation, also known as bra-ket notation:

- A ket $|x\rangle$ denotes a vector, which represents a state of a quantum system.
- A bra $\langle f|$ denotes a linear function that maps each vector to a complex number.
- The action of the linear function $\langle f|$ on a vector $|x\rangle$ is written as $\langle f|x\rangle$.

They are related as follows:

$$|x\rangle = \begin{pmatrix} x_1 \\ x_2 \\ \dots \\ x_n \end{pmatrix} \qquad \langle x| = (|x\rangle^*)^T = (x_1^*, \quad x_2^*, \quad \dots \quad x_n^*)$$

Inner product of two vectors

An inner product over a vector space V over $\mathbb{C}$ is a complex function $(\cdot, \cdot)$ of two vectors that returns a scalar, and which satisfies the following:

- For any x in V, $(x, x) \geq 0$. Moreover $(x, x) = 0$ if and only if $x = 0$.
- For any x, y, z in V, $(\alpha x + \beta y, z) = \alpha(x, z) + \beta(y, z)$.
- For any x, y in V, $(x, y) = (y, x)^*$.

On $\mathbb{C}^n$ the standard Hermitian inner product is: $(x, y) \overset{\text{def}}{=} \sum_{i=1}^{n} x_i^* y_i$.

Using Dirac's notation, the inner product of vectors $|x\rangle$ and $|y\rangle$ is denoted $\langle x|y\rangle$ and is the same as the result of applying the bra $\langle x|$ to the ket $|y\rangle$:

$$|x\rangle = \begin{pmatrix} x_1 \\ x_2 \\ \dots \\ x_n \end{pmatrix} \quad |y\rangle = \begin{pmatrix} y_1 \\ y_2 \\ \dots \\ y_n \end{pmatrix}$$

$$\langle x|y\rangle \stackrel{\text{def}}{=} (x_1^*, \quad x_2^*, \quad \dots \quad x_n^*) \begin{pmatrix} y_1 \\ y_2 \\ \dots \\ y_n \end{pmatrix} = x_1^* y_1 + x_2^* y_2 + \dots + x_n^* y_n$$

The Python numpy.vdot function returns the Hermitian inner product of two vectors:

```
array_to_latex(x, prefix='x = ')
```

$$x = \begin{bmatrix} i \\ 2 \end{bmatrix}$$

```
array_to_latex(y, prefix='y = ')
```

$$y = \begin{bmatrix} -1 \\ 2i \end{bmatrix}$$

```
print("np.vdot(x, y) = ", np.vdot(x, y)
```

$$\mathrm{np.\,vdot(x, y)} = 5j$$

Norm of a vector

The inner product yields a norm defined by $\|x\| = \langle x|x\rangle^{\frac{1}{2}}$. In addition to the triangle inequality $\|x + y\| \leq \|x\| + \|y\|$, the norm also satisfies the Schwarz inequality $\langle x|y\rangle \leq \|x\| \|y\|$. The vector norm or the vector's magnitude is commonly known as the length of the vector.

The Python numpy.linalg.norm function returns the norm of a vector:

```
print("Norm of vector x: {:.3f}".format(np.linalg.norm(x)))
```

$$\mathrm{Norm\ of\ vector\ x:\ 2.236}$$

Hilbert space

An inner-product space V is a Hilbert space if it is complete under the induced norm, that is, if every Cauchy sequence converges: for every sequence $\{x_n\}$ with $x_n \in V$ such that $\lim_{n,m \to \infty} \|x_n - x_m\| = 0$ there is an x in V with $\lim_{n \to \infty} \|x_n - x\| = 0$. This property allows the technique of calculus to be used.

Matrix multiplication with a vector

The @ operator introduced in Python 3.5 implements matrix multiplication with a vector:

```
A = np.array([[1, 2],
              [3, 1j]])
array_to_latex(A, prefix='A = ')
```

$$A = \begin{bmatrix} 1 & 2 \\ 3 & i \end{bmatrix}$$

```
a = np.array([[1],
              [1]])
array_to_latex(a, prefix='a = ')
```

$$a = \begin{bmatrix} 1 \\ 1 \end{bmatrix}$$

```
array_to_latex(A@ x, prefix='A@ x = ')
```

$$A@a = \begin{bmatrix} 3 \\ 3 + i \end{bmatrix}$$

Matrix addition

The addition of two matrices of the same shape is achieved by adding the corresponding entries together:

$$(A + B)_{j,k} = A_{j,k} + B_{j,k}$$

```
A = np.array([[1, 0],
              [0, 1j]])
array_to_latex(A, prefix='A = ')
```

$$A = \begin{bmatrix} 1 & 0 \\ 0 & i \end{bmatrix}$$

```
B = np.array([[0, 1],
              [1j, 0]])
array_to_latex(B, prefix='B = ')
```

$$B = \begin{bmatrix} 0 & 1 \\ i & 0 \end{bmatrix}$$

```
array_to_latex(A+B, prefix='A+B = ')
```

$$A + B = \begin{bmatrix} 1 & 1 \\ i & i \end{bmatrix}$$

Matrix multiplication

Let A be an m-by-n matrix and B an n-by-p matrix, then the product AB is the m-by-p matrix defined as follows:

$$AB_{j,k} = \sum_{r=1}^{n} A_{j,r} B_{r,k}$$

The @ operator introduced in Python 3.5 implements matrix multiplication:

```
A = np.array([[1, 0],
              [0, 1j]])
array_to_latex(A, prefix='A = ')
```

$$A = \begin{bmatrix} 1 & 0 \\ 0 & i \end{bmatrix}$$

```
B = np.array([[1, 1, 1j],
              [1, -1, 0]])
array_to_latex(B, prefix='B = ')
```

$$B = \begin{bmatrix} 1 & 1 & i \\ 1 & -1 & 0 \end{bmatrix}$$

```
array_to_latex(A@B, prefix='A@B = ')
```

$$A@B = \begin{bmatrix} 1 & 1 & i \\ i & -i & 0 \end{bmatrix}$$

Matrix inverse

The inverse of some matrix A when it exists is denoted as A^{-1} is a matrix such that $A^{-1}A = AA^{-1} = \mathbb{1}$ where $\mathbb{1}$ is the identity matrix, for any matrix $A : A\mathbb{1} = \mathbb{1}A = \mathbb{1}$. The numpy.linalg.inv function computes the multiplicative matrix inverse of a matrix:

```
from numpy.linalg import inv
a = np.array([[1., 2.], [3., 4.]])
array_to_latex(A, prefix='A =')
```

$$A = \begin{bmatrix} 1 & 2 \\ 3 & 4 \end{bmatrix}$$

```
array_to_latex(inv(A), prefix='A^{-1} = ')
```

$$A^{-1} = \begin{bmatrix} -2 & 1 \\ \dfrac{3}{2} & -\dfrac{1}{2} \end{bmatrix}$$

Tensor product

Given vector spaces U of dimension m and V of dimension n over $\mathbb{C}$ the tensor product $U \otimes V$ is another vector space W of dimension mn over $\mathbb{C}$.

$\forall \alpha, \beta \in \mathbb{C}$, $\forall u_1, u_2 \in U$, $\forall v_1, v_2 \in V$ and $\forall A, B$ linear maps on U and $\forall C, D$ linear maps on V:

Bilinearity

$$(u_1 + u_2) \otimes v_1 = u_1 \otimes v_1 + u_2 \otimes v_1$$
$$u_1 \otimes (v_1 + v_2) = u_1 \otimes v_1 + u_1 \otimes v_2$$
$$(\alpha u_1) \otimes (\beta v_1) = \alpha \beta (u_1 \otimes v_1)$$

Associativity

$$A \otimes (B \otimes C) = (A \otimes B) \otimes C$$
$$v_1 \otimes (u_1 \otimes u_2) = (v_1 \otimes u_1) \otimes u_2$$

Linear maps properties

$$(A \otimes C)(B \otimes D) = AB \otimes CD$$
$$(A \otimes C)(u_1 \otimes v_1) = Au_1 \otimes Cv_1$$

If the inner-product space W is the tensor product of two inner-product spaces U, V, then for each pair of vectors $|u\rangle \in U$, $|v\rangle \in V$ there is an associated tensor product $|u\rangle \otimes |v\rangle$ in W.

In Dirac's notation, we denote the tensor product $|u\rangle \otimes |v\rangle$ as $|u\rangle|v\rangle$ or $|uv\rangle$.

The inner product of $|u_1\rangle \otimes |v_1\rangle$ and $|u_2\rangle \otimes |v_2\rangle$ is $\langle u_1|u_2\rangle . \langle v_1|v_2\rangle$.

Kronecker product or tensor product of matrices or vectors

The Kronecker product or tensor product denoted as $\otimes$ of two matrices is a composite matrix made of blocks of the second matrix scaled by the first. Let A be an m-by-n matrix and B an p-by-q matrix, then the Kronecker product $A \otimes B$ is the pm-by-qn block matrix:

$$A \otimes B = \begin{pmatrix} a_{11}\,B & \dots & a_{1n}\,B \\ \dots & \dots & \dots \\ a_{m1}\,B & \dots & a_{mn}\,B \end{pmatrix}$$

The Python `numpy.kron` function implements the Kronecker product:

```
A = np.array([[1,2],
              [3, 4]])
array_to_latex(A, prefix='A =')
```

$$A = \begin{bmatrix} 1 & 2 \\ 3 & 4 \end{bmatrix}$$

```
B = np.array([[0, 5],
              [6, 7]])
array_to_latex(B, prefix='B =')
```

$$B = \begin{bmatrix} 0 & 5 \\ 6 & 7 \end{bmatrix}$$

```
C = np.kron(A,B)
array_to_latex(C, prefix='A \otimes B =')
```

$$A \otimes B = \begin{bmatrix} 0 & 5 & 0 & 10 \\ 6 & 7 & 12 & 14 \\ 0 & 15 & 0 & 20 \\ 18 & 21 & 24 & 28 \end{bmatrix}$$

Kronecker sum

The Kronecker sum of any two square matrices, A n-by-n and B m-by-m, noted $A \oplus B$ is defined by:

$$A \oplus B = A \otimes \mathbb{1}_m + \mathbb{1}_n \otimes B$$

where $\mathbb{1}_m$ is the identity matrix or order m and $\mathbb{1}_n$ is the identity matrix of order n.

Outer product

The outer product of a ket $|x\rangle$ and a bra $\langle y|$ is the rank-one operator $|x\rangle\langle y|$ with the rule:

$$(|x\rangle\langle y|)(z) = \langle y|z\rangle|x\rangle$$

For a finite-dimensional vector space, the outer product is a simple matrix multiplication:

$$|x\rangle\langle y| \overset{\text{def}}{=} \begin{pmatrix} x_1 \\ x_2 \\ \dots \\ x_n \end{pmatrix} (y_1^*, \quad y_2^*, \quad \dots \quad y_n^*) = \begin{pmatrix} x_1 y_1^* & x_1 y_2^* & \dots & x_1 y_n^* \\ x_2 y_1^* & x_2 y_2^* & \dots & x_2 y_n^* \\ \dots & \dots & \dots & \dots \\ x_n y_1^* & x_n y_2^* & \dots & x_n y_n^* \end{pmatrix}$$

The Python `numpy.outer` function implements the outer product:

```
array_to_latex(x, prefix='x = ')
```

$$x = \begin{bmatrix} i \\ 2 \end{bmatrix}$$

```
array_to_latex(y, prefix='y = ')
```

$$y = \begin{bmatrix} -1 \\ 2i \end{bmatrix}$$

```
array_to_latex(np.outer(x, y), prefix='np.outer(x, y) = ')
```

$$np.outer(x, y) = \begin{bmatrix} -i & -2 \\ -2 & 4i \end{bmatrix}$$

Writing matrices as a sum of outer products

Any matrix can be written in terms of outer products. For instance, for a 2 x 2 matrix:

$$|0\rangle\langle0| = \begin{pmatrix} 1 \\ 0 \end{pmatrix} \begin{pmatrix} 1 & 0 \end{pmatrix} = \begin{pmatrix} 1 & 0 \\ 0 & 0 \end{pmatrix} \qquad |1\rangle\langle1| = \begin{pmatrix} 0 \\ 1 \end{pmatrix} \begin{pmatrix} 0 & 1 \end{pmatrix} = \begin{pmatrix} 0 & 0 \\ 0 & 1 \end{pmatrix}$$

$$|0\rangle\langle1| = \begin{pmatrix} 1 \\ 0 \end{pmatrix} \begin{pmatrix} 0 & 1 \end{pmatrix} = \begin{pmatrix} 0 & 1 \\ 0 & 0 \end{pmatrix} \qquad |1\rangle\langle0| = \begin{pmatrix} 0 \\ 1 \end{pmatrix} \begin{pmatrix} 1 & 0 \end{pmatrix} = \begin{pmatrix} 0 & 0 \\ 1 & 0 \end{pmatrix}$$

$$M = \begin{pmatrix} m_{0,0} & m_{0,1} \\ m_{1,0} & m_{1,1} \end{pmatrix} = m_{0,0}|0\rangle\langle0| + m_{0,1}|0\rangle\langle1| + m_{1,0}|1\rangle\langle0| + m_{1,1}|1\rangle\langle1|$$

Hermitian operator

The complex conjugate transpose of some vector a or matrix A is often denoted as $a^\dagger$, $A^\dagger$ in quantum mechanics. The symbol, $\dagger$, is called the dagger. $A^\dagger$ is called the adjoint or Hermitian conjugate of A.

A linear operator U is called Hermitian or self-adjoint if it is its own adjoint: $U^\dagger = U$.

The spectral theorem says that if U is Hermitian then it must have a set of orthonormal eigenvectors

$$\{|e_i\rangle \; ; i \in [1, N], \qquad \langle e_i|e_j\rangle = \delta_{ji}\}$$

where $\delta_{ji} = \begin{cases} 0, & i \neq j \\ 1, & i = j \end{cases}$ with real eigenvalues λ_i, $U|e_i\rangle = \lambda_i|e_i\rangle$, and N is the number of eigenvectors, or also is the dimension of the Hilbert space. Hermitian operators have a unique spectral representation in terms of the set of eigenvalues $\{\lambda_i\}$ and the corresponding eigenvectors $|e_i\rangle$:

$$U = \sum_{i=1}^{N} \lambda_i|e_i\rangle\langle e_i|$$

Unitary operator

A linear operator U is called unitary if its adjoint exists and satisfies $U^\dagger U = UU^\dagger = \mathbb{1}$ where $\mathbb{1}$ is the identity matrix, which by definition leaves any vector it is multiplied with unchanged.

Unitary operators preserve inner products:

$$\langle Ux|Uy\rangle = \langle x|U^\dagger U|y\rangle = \langle x|\mathbb{1}|y\rangle = \langle x|y\rangle$$

Hence unitary operators also preserve the norm commonly known as the length of quantum states:

$$\|Ux\| = \langle Ux|Ux\rangle^{\frac{1}{2}} = \langle x|x\rangle^{\frac{1}{2}} = \|x\|$$

For any unitary matrix U, any eigenvectors $|x\rangle$ and $|y\rangle$ and their eigenvalues λ_x and λ_y, $U|x\rangle = \lambda_x|x\rangle$ and $U|y\rangle = \lambda_y|y\rangle$, the eigenvalues λ_x and λ_y have the form $e^{i\theta}$ and if $\lambda_x \neq \lambda_y$ then the eigenvectors $|x\rangle$ and $|y\rangle$ are orthogonal: $\langle x|y\rangle = 0$.

It is useful to note that since for any θ, $\left|e^{i\theta}\right| = 1$:

$$\left|e^{ia} + e^{ib}\right| = \left|e^{\frac{i(a+b)}{2}}\left(e^{\frac{i(a-b)}{2}} + e^{-\frac{i(a-b)}{2}}\right)\right| = \left|e^{\frac{i(a+b)}{2}}\right|\left|e^{\frac{i(a-b)}{2}} + e^{-\frac{i(a-b)}{2}}\right| = \left|e^{\frac{i(a-b)}{2}} + e^{-\frac{i(a-b)}{2}}\right|$$

Density matrix

Any quantum state, either **mixed** or **pure**, can be described by a **density matrix** (ρ), which is a normalized positive Hermitian operator where $\rho = \rho^{\dagger}$. According to the spectral theorem, there exists an orthonormal basis, defined in *Section 2.3.1, Hermitian operator*, such that the density is the sum of all eigenvalues (N):

$$\rho = \sum_{i=1}^{N} \lambda_i|e_i\rangle\langle e_i|$$

where i ranges from 1 to N, λ_i are positive or null eigenvalues ($\lambda_i \geq 0$), and the sum of eigenvalues is the trace operation (tr) of the density matrix and is equal to 1:

$$tr(\rho) = \sum_{i=1}^{N} \lambda_i = 1$$

For example, when the density is $\rho = \begin{pmatrix} \rho_{0,0} & \rho_{0,1} \\ \rho_{1,0} & \rho_{1,1} \end{pmatrix}$, with $\rho = \rho^{\dagger}$, the trace of the density is:

$$tr(\rho) = \rho_{0,0} + \rho_{1,1} = 1$$

Here are some examples of the density matrices of pure quantum states:

$$\begin{pmatrix} 1 & 0 \\ 0 & 0 \end{pmatrix} = \begin{pmatrix} 1 \\ 0 \end{pmatrix} \begin{pmatrix} 1 & 0 \end{pmatrix} = |0\rangle\langle 0|$$

$$\begin{pmatrix} 0 & 0 \\ 0 & 1 \end{pmatrix} = \begin{pmatrix} 0 \\ 1 \end{pmatrix} \begin{pmatrix} 0 & 1 \end{pmatrix} = |1\rangle\langle 1|$$

$$\frac{1}{2}\begin{pmatrix} 1 & -1 \\ -1 & 1 \end{pmatrix} = \frac{1}{2}(|0\rangle - |1\rangle)(|0\rangle - |1\rangle)$$

The density matrix of a mixed quantum state consisting of a statistical ensemble of n pure quantum states $\{|x_i\rangle ; i \in [1,n]\}$, each with a classical probability of occurrence p_i, is defined as:

$$\rho = \sum_{i=1}^{n} p_i |x_i\rangle\langle x_i|$$

where every p_i is positive or null and their sum is equal to one:

$$tr(\rho) = \sum_{i=1}^{n} p_i = 1$$

We summarize the difference between pure states and mixed states in Figure AA.1 which is the same as *Figure 2.20*.

Density matrix of a pure state $|x\rangle$	Density matrix of a mixed state $\{|x_i\rangle ; i \in [1,n]\}$
$\rho = |x\rangle\langle x|$	$\rho = \sum_{i=1}^{n} p_i |x_i\rangle\langle x_i|$
$tr(\rho^2) = \sum_{i=1}^{N} \lambda_i^2 = 1$	$tr(\rho^2) = \sum_{i=1}^{N} \lambda_i^2 < 1$

Figure AA.1 – Density matrix of pure and mixed quantum states

Pauli matrices

There are three Pauli matrices (σ_x, σ_y and σ_z):

$$\sigma_x = \begin{pmatrix} 0 & 1 \\ 1 & 0 \end{pmatrix}, \quad \sigma_y = \begin{pmatrix} 0 & -i \\ i & 0 \end{pmatrix}, \quad \sigma_z = \begin{pmatrix} 1 & 0 \\ 0 & -1 \end{pmatrix}$$

which are Hermitian and unitary making the square of each equal to the (2×2) identity matrix:

$$\sigma_x^2 = \sigma_y^2 = \sigma_z^2 = \begin{pmatrix} 1 & 0 \\ 0 & 1 \end{pmatrix}$$

Each of the Pauli matrices is equal to its inverse:

$$\sigma_x = \sigma_x^{-1}$$
$$\sigma_y = \sigma_y^{-1}$$
$$\sigma_z = \sigma_z^{-1}$$

We summarize the Pauli matrices and the operations on a qubit that yields the associated eigenvectors in the following table:

Pauli matrix	Eigenvector
$\sigma_x = \begin{pmatrix} 0 & 1 \\ 1 & 0 \end{pmatrix}$	$\lvert + \rangle = \frac{1}{\sqrt{2}} \lvert 0 \rangle + \frac{1}{\sqrt{2}} \lvert 1 \rangle$
Sign basis: $\{\lvert + \rangle, \lvert - \rangle\}$	$\lvert - \rangle = \frac{1}{\sqrt{2}} \lvert 0 \rangle - \frac{1}{\sqrt{2}} \lvert 1 \rangle$
$\sigma_y = \begin{pmatrix} 0 & -i \\ i & 0 \end{pmatrix}$	$\lvert i+ \rangle = \frac{1}{\sqrt{2}} \lvert 0 \rangle + \frac{i}{\sqrt{2}} \lvert 1 \rangle$
Complex basis: $\{\lvert i+ \rangle, \lvert i- \rangle\}$	$\lvert i- \rangle = \frac{1}{\sqrt{2}} \lvert 0 \rangle - \frac{i}{\sqrt{2}} \lvert 1 \rangle$
$\sigma_z = \begin{pmatrix} 1 & 0 \\ 0 & -1 \end{pmatrix}$	$\lvert 0 \rangle$
Standard basis: $\{\lvert 0 \rangle, \lvert 1 \rangle\}$	$\lvert 1 \rangle$

Decomposing a matrix into the weighted sum of the tensor product of Pauli matrices

It can be shown that any matrix can be decomposed into the weighted sum of the tensor product of the identity matrix and the Pauli matrices $P_i = \otimes_j^N \sigma_{i,j}$ where $\sigma_{i,j} \in \{\mathbb{1}, \sigma_x, \sigma_y, \sigma_z\}$ with weights h_i and N qubits:

$$M = \sum_{i=1}^{n} h_i \otimes_j^N \sigma_{i,j}$$

For Hermitian matrices, all weights h_i are real.

We provide a proof for any 2x2 matrix, $M = \begin{pmatrix} m_{0,0} & m_{0,1} \\ m_{1,0} & m_{1,1} \end{pmatrix}$.

$$\sigma_z\, \sigma_x = \begin{pmatrix} 1 & 0 \\ 0 & -1 \end{pmatrix}\begin{pmatrix} 0 & 1 \\ 1 & 0 \end{pmatrix} = \begin{pmatrix} 0 & 1 \\ -1 & 0 \end{pmatrix} = i\sigma_y$$

$$\sigma_x\sigma_z = \begin{pmatrix} 0 & 1 \\ 1 & 0 \end{pmatrix}\begin{pmatrix} 1 & 0 \\ 0 & -1 \end{pmatrix} = \begin{pmatrix} 0 & -1 \\ 1 & 0 \end{pmatrix} = -i\sigma_y$$

$$\frac{\mathbb{1} + \sigma_z}{2} = \frac{1}{2}\left(\begin{pmatrix} 1 & 0 \\ 0 & 1 \end{pmatrix} + \begin{pmatrix} 1 & 0 \\ 0 & -1 \end{pmatrix}\right) = \begin{pmatrix} 1 & 0 \\ 0 & 0 \end{pmatrix} = |0\rangle\langle0|$$

$$\frac{\mathbb{1} - \sigma_z}{2} = \frac{1}{2}\left(\begin{pmatrix} 1 & 0 \\ 0 & 1 \end{pmatrix} - \begin{pmatrix} 1 & 0 \\ 0 & -1 \end{pmatrix}\right) = \begin{pmatrix} 0 & 0 \\ 0 & 1 \end{pmatrix} = |1\rangle\langle1|$$

Since $\sigma_x|0\rangle = |1\rangle$ hence $\langle1| = \langle0|\sigma_x$ we have:

$$|0\rangle\langle1| = |0\rangle\langle0|\sigma_x = \frac{\mathbb{1} + \sigma_z}{2}\sigma_x = \frac{\sigma_x + i\sigma_y}{2}$$

$$|1\rangle\langle0| = \sigma_x|0\rangle\langle0| = \sigma_x\frac{\mathbb{1} + \sigma_z}{2} = \frac{\sigma_x - i\sigma_y}{2}$$

Starting from the decomposition of a 2x2 matrix as a sum of outer products:

$$M = \begin{pmatrix} m_{0,0} & m_{0,1} \\ m_{1,0} & m_{1,1} \end{pmatrix} = m_{0,0}|0\rangle\langle0| + m_{0,1}|0\rangle\langle1| + m_{1,0}|1\rangle\langle0| + m_{1,1}|1\rangle\langle1|$$

We can then write:

$$M = m_{0,0}\frac{\mathbb{1} + \sigma_z}{2} + m_{0,1}\frac{\sigma_x + i\sigma_y}{2} + m_{1,0}\frac{\sigma_x - i\sigma_y}{2} + m_{1,1}\frac{\mathbb{1} - \sigma_z}{2}$$

$$M = \frac{m_{0,0} + m_{1,1}}{2}\mathbb{1} + \frac{m_{0,1} + m_{1,0}}{2}\sigma_x + i\frac{m_{0,1} - m_{1,0}}{2}\sigma_y + \frac{m_{0,0} - m_{1,1}}{2}\sigma_z$$

Anti-commutator

An operation of two operators A, B defined as: $\{A, B\} \stackrel{\text{def}}{=} AB + BA$.

Anti-commutation

A set of fermionic annihilation operators $\{\hat{a}_i\}_{i=0}^{N-1}$ and creation operators $\left\{\hat{a}_j^\dagger\right\}_{j=0}^{N-1}$ that act on local electron modes can be defined that satisfy the following anti-commutation relations:

$$\{\hat{a}_i, \hat{a}_j^\dagger\} = \hat{a}_i^\dagger \hat{a}_j + \hat{a}_j \hat{a}_i^\dagger = \delta_{ij} = \begin{cases} 0, & i \neq j \\ 1, & i = j \end{cases}$$

$$\{a_i^\dagger, a_j^\dagger\} = \{\hat{a}_i, \hat{a}_j\} = 0$$

Commutator

An operation of two operators A, B defined as: $[A, B] \overset{\text{def}}{=} AB - BA$. For any operators A and B, $[A, B] = 0$ if and only if A and B commute. It can be shown that if a quantum system has two simultaneously physically observable quantities, then the Hermitian operators that represent them must commute. For any operators A, B and C we have the following relations, which are useful for calculating commutators:

$$[A, A] = 0$$
$$[A, B] + [B, A] = 0$$
$$[A, B + C] = [A, B] + [A, C]$$
$$[A + B, C] = [A, C] + [B, C]$$
$$[A, BC] = [A, B]C + B[A, C]$$
$$[AB, C] = [A, C]B + A[B, C]$$
$$\left[A, [B, C]\right] + \left[C, [A, B]\right] + \left[B, [C, A]\right] = 0$$

Fermion, fermionic, electron annihilation operator

A mathematical operation that allows us to represent excitations or transitions of quasi-particles. An excitation requires the initial state to be at a lower energy level than the final state.

An operator $\hat{a}_i$ that lowers by one unit the number of particles sitting in the ith fermionic orbital:

$$\hat{a}_i|\ldots m_i \ldots\rangle = m_i\, (-1)^{\Sigma_{j<i}\, m_j}|\ldots (m-1)_i \ldots\rangle$$

where:

m_i and $(m - 1)_i$ are the number of particles sitting in the ith fermionic orbital.

m_i is a pre-factor that annihilates the state in the Slater determinant if there is no electron in the ith fermionic orbital, that is, if $m_i = 0$.

The phase factor $(-1)^{\Sigma_{j<i}m_j}$ keeps the anti-symmetric properties of the whole superposition of states.

Fermion, fermionic, electron creation operator

A mathematical operation that allows us to represent disexcitement (de-exitation) or transitions of quasi-particles. A de-excitation requires the initial state to be at a higher energy level than the final state.

An operator $\hat{a}_i^\dagger$ that raises by one unit the number of particles sitting in the ith fermionic orbital:

$$\hat{a}_i^\dagger | ... \, m_i \, ...\rangle = (1 - m_i)\,(-1)^{\Sigma_{j<i}m_j}| ... \, (m+1)_i \, ...\rangle$$

where:

m_i and $(m+1)_i$ are the number of particles sitting in the ith fermionic orbital.

$(1 - m_i)$ is a pre-factor that annihilates the state if we had an electron in the ith fermionic orbital, that is, if $m_i = 1$.

The phase factor $(-1)^{\Sigma_{j<i}m_j}$ keeps the anti-symmetric properties of the whole superposition of states.

Fermion, fermionic, electron excitation operator

An operator $\hat{a}_i^\dagger \hat{a}_j$ that excites an electron from the occupied spin orbital $\psi_j(\mathbf{r}_p)\chi_j(\mathbf{s}_p)$ into the unoccupied orbital $\psi_i(\mathbf{r}_p)\chi_i(\mathbf{s}_p)$.

Total wave function

Describes the physical behavior of a system and is represented by the capital Greek letter Psi: Ψ_{total}. It contains all the information of a quantum system including complex numbers ($z = a + ib$) as parameters. In general, Ψ_{total} is a function of all the particles in the system $\{1, ..., i, ..., N\}$, where the total number of particles is N. Furthermore, Ψ_{total} includes the spatial position of each particle ($\mathbf{r}_i = \{x_i, y_i, z_i\}$), the spin directional coordinates of each particle ($\mathbf{s}_i = \{s_{x_i}, s_{y_i}, s_{z_i}\}$), and time ($t$):

$$\Psi_{total}(\mathbf{r}, \mathbf{s}, t)$$

where r and s are vectors of single particle coordinates:

$$r = \{r_1, \dots, r_i, \dots, r_N\}$$
$$s = \{s_1, \dots, s_i, \dots, s_N\}$$

The total wave function for a one particle system is a product of a spatial $\psi(r_1)$, spin $\chi(s_1)$, and time $f(t)$ functions:

$$\Psi_{total}(r_1, s_1, t) = \psi(r_1) * \chi(s_1) * f(t)$$

References

[Micr_Algebra] Linear algebra, QuantumKatas/tutorials/LinearAlgebra/: `https://github.com/microsoft/QuantumKatas/tree/main/tutorials/LinearAlgebra`

[Micr_Complex] Complex arithmetic, QuantumKatas/tutorials/ComplexArithmetic/: `https://github.com/microsoft/QuantumKatas/tree/main/tutorials/ComplexArithmetic`

[NumPy] NumPy: the absolute basics for beginners: `https://numpy.org/doc/stable/user/absolute_beginners.html`

[Qiskit] Qiskit: `https://qiskit.org/`

[Qiskit_Alg] Linear Algebra, Qiskit: `https://qiskit.org/textbook/ch-appendix/linear_algebra.html`

[SymPy] SymPy, A Python library for symbolic mathematics: `https://www.sympy.org/en/index.html`

Appendix B
Leveraging Jupyter Notebooks on the Cloud

In this appendix, we will cover the following topics:

- Jupyter Notebook
- Google Colaboratory
- IBM Quantum Lab
- Companion Jupyter notebooks

Jupyter Notebook

The Jupyter Notebook is a free web application for creating and sharing computational documents that combine executable code with narrative text in Markdown format [Jupyter_0]. It offers a simple, streamlined, document-centric experience. Project Jupyter is a non-profit, open-source project.

Google Colaboratory

Google Colaboratory (or Colab for short) is a free Jupyter Notebook environment that runs entirely in the cloud and provides shared online instances of Jupyter notebooks without having to download or install any software [Colab_0] [Colab_1]. You just need to have a working Gmail account to save and access Google Colab Jupyter notebooks.

IBM Quantum Lab

IBM Quantum Lab is a cloud-enabled Jupyter notebook environment that requires no installation [IBM_QLab0] [IBM_QLab1]. IBM Quantum Composer is a graphical quantum programming tool that lets you drag and drop operations to build quantum circuits and run them on real quantum hardware or simulators [IBM_comp1] [IBM_comp2]. It allows public and free access to cloud-based quantum computing services provided by IBM. In Quantum Lab, you can write scripts that combine Qiskit code, equations, visualizations, and narrative text in a customized Jupyter Notebook environment.

Companion Jupyter notebooks

We provide a repository on GitHub of the companion Jupyter notebooks of the book here: `https://github.com/PacktPublishing/Quantum-Chemistry-and-Computing-for-the-Curious`. These companion notebooks automatically install the relevant list of libraries, as follows:

- **Numerical Python (NumPy)** [NumPy], an open-source Python library that is used in almost every field of science and engineering.

- Qiskit [Qiskit], an open-source SDK for working with quantum computers at the level of pulses, circuits, and application modules.

- Qiskit visualization support, to enable use of visualization functionality and Jupyter notebooks.

- Qiskit Nature [Qiskit_Nature] [Qiskit_Nat_0], a unique platform to bridge the gap between natural sciences and quantum simulations.

- **Python-based Simulations of Chemistry Framework (PySCF)** [PySCF] is an open-source collection of electronic structure modules powered by Python.

- **Quantum Toolbox in Python (QuTiP)** [QuTiP] is designed to be a general framework for solving quantum mechanics problems such as systems composed of few-level quantum systems and harmonic oscillators.

- **Atomic Simulation Environment (ASE)** [ASE_0], a set of tools and Python modules for setting up, manipulating, running, visualizing, and analyzing atomistic simulations. The code is freely available under the GNU LGPL license.

- PyQMC [PyQMC], a Python module that implements real-space quantum Monte Carlo techniques. It is primarily meant to interoperate with PySCF.

- h5py [H5py] package, a Pythonic interface to the HDF5 binary data format.

- SciPy [SciPy_0], a free and open-source Python library used for scientific computing and technical computing. SciPy provides algorithms for optimization, integration, interpolation, eigenvalue problems, algebraic equations, differential equations, statistics, and many other classes of problems.

- SymPy [SymPy], a Python library for symbolic mathematics.

All companion Jupyter notebooks have been successfully run in both Google Colab and Quantum Lab environments.

The companion Jupyter notebook of *Chapter 6, Beyond Born-Oppenheimer* does not include the installation of the Psi4, open-source software for high-throughput quantum chemistry [Psi4_0] that we used to perform a simple calculation of the vibrational frequency analysis of the carbon dioxide (CO_2) molecule. We refer the reader interested in installing this package to the "Get Started with Psi4" [Psi4_1] documentation and to the article Ref. [Psi4_3].

References

[ASE_0] Atomic Simulation Environment (ASE), `https://wiki.fysik.dtu.dk/ase/index.html`

[Colab_0] Welcome to Colaboratory, Google Colab FAQ, `https://research.google.com/colaboratory/faq.html`

[Colab_1] Welcome to Colaboratory, `https://colab.research.google.com/`

[H5py] Quick Start Guide, `https://docs.h5py.org/en/stable/quick.html`

[IBM_QLab0] IBM Quantum Lab, `https://quantum-computing.ibm.com/lab`

[IBM_QLab1] Welcome to Quantum Lab, `https://quantum-computing.ibm.com/lab/docs/iql/`

[IBM_comp1] Welcome to IBM Quantum Composer, `https://quantum-computing.ibm.com/composer/docs/iqx/`

[IBM_comp2] IBM Quantum Composer, `https://quantum-computing.ibm.com/composer/files/new`

[Jupyter_0] Jupyter, `https://jupyter.org/`

[NumPy] NumPy: the absolute basics for beginners, `https://numpy.org/doc/stable/user/absolute_beginners.html`

[Psi4_0] Psi4 manual master index, `https://psicode.org/psi4manual/master/index.html`

[Psi4_1] Get Started with PSI4, `https://psicode.org/installs/v15/`

[Psi4_3] Smith DGA, Burns LA, Simmonett AC, Parrish RM, Schieber MC, Galvelis R, Kraus P, Kruse H, Di Remigio R, Alenaizan A, James AM, Lehtola S, Misiewicz JP, Scheurer M, Shaw RA, Schriber JB, Xie Y, Glick ZL, Sirianni DA, O'Brien JS, Waldrop JM, Kumar A, Hohenstein EG, Pritchard BP, Brooks BR, Schaefer HF 3rd, Sokolov AY, Patkowski K, DePrince AE 3rd, Bozkaya U, King RA, Evangelista FA, Turney JM, Crawford TD, Sherrill CD, Psi4 1.4: Open-source software for high-throughput quantum chemistry, J Chem Phys. 2020 May 14;152(18):184108. doi: 10.1063/5.0006002. PMID: 32414239; PMCID: PMC7228781, `https://www.ncbi.nlm.nih.gov/pmc/articles/PMC7228781/pdf/JCPSA6-000152-184108_1.pdf`

[PyQMC] PyQMC, a Python module that implements real-space quantum Monte Carlo techniques, `https://github.com/WagnerGroup/pyqmc`

[PySCF] The Python-based Simulations of Chemistry Framework (PySCF), `https://pyscf.org/`

[Qiskit] Qiskit, `https://qiskit.org/`

[Qiskit_Nat_0] Qiskit_Nature, `https://github.com/Qiskit/qiskit-nature/blob/main/README.md`

[Qiskit_Nature] Introducing Qiskit Nature, Qiskit, Medium, April 6, 2021, `https://medium.com/qiskit/introducing-qiskit-nature-cb9e588bb004`

[QuTiP] QuTiP, Plotting on the Bloch Sphere, `https://qutip.org/docs/latest/guide/guide-bloch.html`

[SciPy_0], SciPy, `https://scipy.org/`

[SymPy] SymPy, A Python library for symbolic mathematics, `https://www.sympy.org/en/index.html`

Appendix C
Trademarks

Atomic Simulation Environment (ASE) is copyright © 2022, ASE-developers.

Google Colab is copyright © 2017, COLAB, LLC.

h5py, a thin, Pythonic wrapper around HDF5, which runs on Python 3 (3.6+), is copyright © 2008, Andrew Collette and contributors.

IBM®, IBM Q Experience®, and Qiskit® are registered trademarks of IBM Corporation.

Linux is the registered trademark of Linus Torvalds in the U.S. and other countries.

Psi4, an open-source quantum chemistry software package, is copyright © 2007-2022, the Psi4 developers.

pyQMC, a Python module that implements real-space quantum Monte Carlo techniques, is copyright © 2019, Lucas K Wagner, pyQMC authors.

The **Python-based Simulations of Chemistry Framework (PySCF)** is copyright © 2014, the PySCF developers.

Python is copyright© 2001-2022 Python Software Foundation.

The Python logo is trademark of the Python Software Foundation: `https://www.python.org/community/logos/`.

Quantum Toolbox in Python (QuTiP) is copyright © 2011-2021 inclusive, QuTiP developers and contributors.

SciPy, an open-source software for mathematics, science, and engineering, is copyright © 2001-2002 Enthought, Inc., and 2003-2022, SciPy developers.

SymPy is copyright © 2021, SymPy Development Team.

The NumPy trademark is registered with the U.S. Patent & Trademark Office (USPTO).

The Jupyter trademark is registered with the U.S. Patent & Trademark Office (USPTO).

Index

322

W

X

Z

Other Books You May Enjoy

If you enjoyed this book, you may be interested in these other books by Packt:

Dancing with Qubits

Robert S. Sutor

ISBN: 9781838827366

- See how quantum computing works, delve into the math behind it, what makes it different, and why it is so powerful with this quantum computing textbook

- Discover the complex, mind-bending mechanics that underpin quantum systems

- Understand the necessary concepts behind classical and quantum computing

- Refresh and extend your grasp of essential mathematics, computing, and quantum theory

- Explore the main applications of quantum computing to the fields of scientific computing, AI, and elsewhere

- Examine a detailed overview of qubits, quantum circuits, and quantum algorithm

Quantum Computing in Practice with Qiskit® and IBM Quantum Experience®

Hassi Norlen

ISBN: 9781838828448

- Visualize a qubit in Python and understand the concept of superposition

- Install a local Qiskit® simulator and connect to actual quantum hardware

- Compose quantum programs at the level of circuits using Qiskit® Terra

- Compare and contrast Noisy Intermediate-Scale Quantum computing (NISQ) and Universal Fault-Tolerant quantum computing using simulators and IBM Quantum® hardware

- Mitigate noise in quantum circuits and systems using Qiskit® Ignis

- Understand the difference between classical and quantum algorithms by implementing Grover's algorithm in Qiskit®

Packt is searching for authors like you

If you're interested in becoming an author for Packt, please visit authors.
packtpub.com and apply today. We have worked with thousands of developers and
tech professionals, just like you, to help them share their insight with the global tech
community. You can make a general application, apply for a specific hot topic that we are
recruiting an author for, or submit your own idea.

Share Your Thoughts

Now you've finished *Quantum Chemistry and Computing for the Curious*, we'd love to hear
your thoughts! Scan the QR code below to go straight to the Amazon review page for this
book and share your feedback or leave a review on the site that you purchased it from.

https://packt.link/r/1-803-24390-2

Your review is important to us and the tech community and will help us make sure we're
delivering excellent quality content.

Made in the USA
Monee, IL
12 July 2022